Mathematics Teaching
The State of the Art

W. Ann Bowker

Mathematics Teaching
The State of the Art

Edited by

Paul Ernest

The Falmer Press
(A member of the Taylor & Francis Group)
New York • Philadelphia • London

UK The Falmer Press, Falmer House, Barcombe, Lewes, East Sussex, BN8 5DL

USA The Falmer Press, Taylor & Francis Inc., 242 Cherry Street, Philadelphia, PA 19106-1906

First published in 1989
Reprinted 1991

Library of Congress Cataloging-in-Publication Data

Mathematics teaching : the state of the art / edited by Paul Ernest.
p. cm.
Includes bibliographies and index.
ISBN 1-85000-460-9. – ISBN 1-85000-461-7 (pbk.)
1. Mathematics – Study and teaching. I. Ernest, Paul.
QA11.M3765 1989
510′.7′1 – dc 19 88-32457
CIP

Typeset in 10½/12 Caledonia by
Imago Publishing Ltd, Thame, Oxon

Printed in Great Britain by
Redwood Press Limited, Melksham, Wiltshire

Contents

Acknowlegments

I would like to acknowledge the contribution of Dr Peter Preece of the University of Exeter, whose suggestion that I edit an issue of *Perspectives* ultimately led to this book. Peter also contributed by his scrupulous proof-reading of the *Perspectives* articles, many of which have been made available to a wider readership by inclusion in this book.

I would also like to thank my wife Jill for all her encouragement, as well as for reading the galley proofs and finding errors that escaped my eyes! Although not involved in education, Jill found the book inspirational, which is a good omen indeed!

Permission has been sought for the use of diagrams, and acknowledgments are due to the National Council of Teachers of Mathematics (for the figure on page 58); the Assessment of Performance Unit (for the figures on pages 59, 65, 66 and 67); Blackie & Sons (for the figure on page 70); NFER-Nelson (for the figures on pages 120 and 121); Foto Inn (for the figure on page 122); Longman (for the figures on pages 124 and 125); Cambridge University Press (for the figures on pages 126, 127 and 128); Bell and Hyman (for the figures on pages 129 and 131); John Murray (for the figures on pages 130, 131 and 132); Stanley Thornes (for the figures on pages 133 and 134) and the Advisory Council for Adult and Continuing Education (for the figures on pages 206 and 207).

Permission has also been sought to reprint three journal articles, and acknowledgments are due to the editor of *Teaching Mathematics and its Applications* (for Chapter 5, from volume 5, number 2, 1986,); to the editor of *Mathematics in School* (for Chapter 3, from volume 17 number 1, 1988) and to the editors of *Mathematics Teaching* (for Chapter 17, from issue number 116, 1986).

Introduction

Paul Ernest

Mathematics teaching is currently in a state of ferment and change. The products of the new information and microchip technologies, particularly the microcomputer, have unlimited potential and transform the teaching of mathematics. Official reports on school mathematics urge the adoption of new styles of teaching. The introduction of new forms of testing, most notably the GCSE examination, requires radically different patterns of assessment and teaching. At the same time research is at last delivering knowledge of children's understanding of mathematics, and providing exciting new perspectives on the teaching of mathematics. At no time in the history of mathematics teaching have there been changes on so broad a front, or so rapid a growth of knowledge as today. This book attempts to offer teachers, educationalists and all other interested persons an overview of these changes, from the perspective of experts and practitioners at the leading edge of development.

However, this book attempts to provide more than just a picture of current practice and innovation. It also offers a state of the art review of research in mathematics education, in an accessible form. Uniquely, it brings together leading researchers of national and international renown, both from the mathematics education community and from the broader educational research community.

The three sections of the book treat innovations in school mathematics, research perspectives on the teaching and learning of mathematics, including that of constructivism, and research on the social context of mathematics, including the issues of gender, race, culture and social and political values. Although not exhaustive, the book covers a broad spectrum of issues and expert opinion. It thus offers a unique but accessible survey of the state of the art in mathematics education, as we enter the 1990s.

Since the focus of the book is the present state of mathematics teaching, with an eye to future prospects, it is appropriate to begin by tracing some of the immediate antecedents of the current British scene.

The Background to the Current British Scene

The roots of the present can be traced back as far as one likes, as Howson (1982) shows. However, the 1970s provide a suitable starting point for the present discussion, for many of the key determinants of the present came into focus then. During the 1970s the following events took place.

The hand-held electronic calculator and the microcomputer had their first impact on education.

The back-to-basics movement emerged, as a blacklash to the modern maths and progressive education of the 1960s.

A concern with standards of achievement, including those in mathematics, was voiced by Prime Minister James Callaghan at the famous Ruskin College speech in 1976, initiating what became known as 'The Great Debate' on education.

The parliamentary Committee of Inquiry into the Teaching of Mathematics was established, chaired by Sir Wilfrid Cockcroft.

The Assessment of Performance Unit was established at the Department of Education and Science to monitor standards of achievement by schoolchildren of 11 and 15 years of age, in a number of areas including mathematics.

Mathematics education as an academic discipline became established and institutionalized, with the emergence of Masters degree programmes, such as at the Polytechnic of the South Bank, and centres of research and development, especially the Shell Centres at Chelsea College, London and the University of Nottingham.

These developments set the scene for developments in the teaching of mathematics in the 1980s.

The most important event for the teaching of mathematics in Britain in the 1980s was the publication of the report of the Cockcroft Committee (1982). This well written endorsement of progressive practice was enthusiastically received at all levels by those involved in education, and beyond in the wider community. The report was sufficiently authoritative to deflect criticism which had been levelled at mathematics teaching and levels of achievement away from a call to return to basics. The report covered a great number of issues, and made many recommendations. Some of the most significant issues treated were the following.

Recognition of the importance and the (then unfulfilled) potential of microcomputers and electronic calculators for the teaching of mathematics;

Acknowledgment that the learning of mathematics involves more than basic knowledge of facts, skills and procedures. It also crucially involves conceptual structures, the general strategies of problem-solving and attitudes to and the appreciation of mathematics;

Acknowledgment that successful learning requires a range of teaching styles at all levels, including discussion, problem-solving, mathematical investigation, appropriate practical work, in addition to the more traditional approaches of exposition, and the practice and reinforcement of skills;

A powerful critique of the traditional forms of school assessment in mathematics at age 16, which for a great many children focused more on failure than on mastery and success. Associated with this is a further critique of the nature and demands of the mathematics curriculum for the below average attaining student;

Recognition of the inadequacy of much of the teaching in schools due to the insufficient training and preparation of many teachers in mathematics, and the use of non-specialists to teach mathematics;

Recognition of the necessity for curriculum leadership in school mathematics at all levels, including the coordination of mathematics in primary and middle schools, and a recognition of the complex demands of this role, and that of the heads of mathematics departments in secondary schools;

Recognition of the importance of mathematics in the world of work and in society in general, tempered with knowledge of many adults' ignorance and fear of mathematics. More generally, there was an acknowledgment of the importance of broader social issues and contexts for the teaching of mathematics in a number of ways, especially with regard to gender: the under-achievement of girls in mathematics.

Following this landmark report, there were important official responses, although the report was only one of the goads to action where there were general educational problems. Whatever their causes, the following official actions significantly affected the teaching of mathematics in the 1980s.

Government agencies funded the provision of computer hardware for all British schools (the Department of Trade and Industry initiative) and the development of good software and in-service materials for microcomputers in schools, especially in mathematics (the Microelectronics Education Programme).

Traditional assessments at 16+ were replaced by the GCSE examination which addressed many of the criticisms of the Cockcroft Report. It heralded a move from norm-referenced to criterion-referenced assessment, and incorporated novel oral and course work modes involving extended pieces of work (projects, mathematical investigations and practical work).

A large number (350) of advisory teachers of mathematics were appointed (the so-called 'Cockcroft missionaries'), funded by the Educational Support Grant scheme, to work with teachers and in classrooms to 'lever up' the standards of mathematics teaching.

New Styles of Teaching

Perhaps the most important development in the teaching of mathematics in the 1980s has been the widespread recognition and adoption of the range of teaching styles endorsed by the Cockcroft Report, in the celebrated paragraph 243.

> **Mathematics teaching at all levels should include opportunities for**
> * exposition by the teacher;
> * discussion between teacher and pupils and between pupils themselves;
> * appropriate practical work;
> * consolidation and practice of fundamental skills and routines;
> * problem solving, including the application of mathematics to everyday situations;
> * investigational work. (Cockcroft, 1982, paragraph 243)

This endorsement gave a new authority to the recommendations of progressive mathematics educators and teacher organizations, such as the radical Association of Teachers of Mathematics. These recommendations were moved to the top of the education agenda by the inclusion of extended pieces of work such as mathematical projects and investigations for assessment, to be mandatory by the year 1991, in the *GCSE National Criteria for Mathematics* (SEC, 1985), and the associated national in-service training campaign. This move was intended to produce assessment-driven innovation in mathematics teaching styles. It was a policy which raised consciousness (and anxiety levels) very rapidly, but affected classroom practice more slowly.

Although the changes in teaching styles have been gradual, with varied rates of adoption, the onus has been placed on teachers as a body to adapt, rather than on a progressive minority to proselytize. The move towards new styles of teaching echoes a worldwide recognition that to fulfil personal and societal needs the goals of mathematics education must include the development of higher level thinking skills, particularly problem-solving. Thus, for example, in the USA the largest mathematics teacher organization in the world, the National Council of Teachers of Mathematics, issued a policy statement which stated as the first of ten recommendations that: 'Problem Solving must be the focus of school mathematics in the 1980's' (NCTM, 1980, p. 1).

The increasingly widespread use of microcomputers has also had an impact on mathematics teaching styles. When children interact with software, the teacher's role naturally becomes more that of facilitator than instructor. Programming in computer languages such as BASIC, LOGO or Prolog also requires the teacher to adopt this role, not least because children's knowledge and expertise often rapidly outstrips the teacher's!

Since the Cockcroft Report, there have been three further official publications concerning the teaching of mathematics: HMI (1985), Low Attainers in Mathematics Project (1987) and National Curriculum Mathematics Working Group (1987). Each of these has endorsed and further elaborated the Cockcroft recommendations concerning the range of styles necessary for effective mathematics teaching. Thus although these recommendations are not fully implemented, the mathematics teaching community in Britain is virtually unanimous in its support for a more problem-solving orientation to mathematics teaching, incorporating discussion, practical and investigative work.

The Fruits of Research

In the 1980s increased research on the teaching of mathematics has at last been bearing fruit, both in terms of an increased knowledge of children's learning of mathematics and in providing new perspectives with which to understand better the teaching of mathematics and its broader context.

Important empirical data on children's understanding of mathematics have been collected, particularly by the Concepts in Secondary Mathematics and Science Project (CSMS), reported in Hart (1981), and the Assessment of Performance Unit (1985), which summarized the results of its research, published from 1979 onwards. Until these research results were published the gap between the intended mathematics curriculum and that actually delivered was not widely appreciated. These projects also alerted mathematics teachers to the importance and widespread nature of children's idiosyncratic strategies and errors in mathematics. Indeed, a follow-up to the CSMS project was named the 'Strategies and Errors in Mathematics Project', and investigated and attempted to remediate children's specific errors in algebra, ratio and fractions (Booth, 1984; Hart, 1984; Kerslake, 1984). The Assessment of Performance Unit, in addition to publishing a great deal of data on children's mathematical achievement, has developed and carried out widespread testing of innovative designs, notably the assessment of attitudes, oral and practical work, problem-solving, and cooperative group work, including pupil-to-pupil discussion.

Other recent research has greatly illuminated the processes of learning mathematics. The applications of approaches from a number of research perspectives within cognitive psychology, including cognitive science (Carpenter *et al.*, 1982; Davis, 1984), information processing theory (Brown and Burton, 1978) and constructivism (Steffe *et al.*, 1983), are providing plausible models of how students learn and perform mathematics. Conceptual frameworks such as these are providing the mathematics teacher with an understanding which is vital for effective teaching. In addition, much is being learnt about aspects of mathematics instruction including teaching approaches and instructional aids (Bell *et al.*, 1983; Good *et al.*, 1983; Christiansen *et al.*, 1985), all of which have the potential to enhance the teaching of mathematics.

The 1980s has seen the growth of research on a further aspect of the teaching of mathematics, namely its social context. This is now considered to be so significant that one whole section of this book is devoted to the area. Some of the key research issues and a brief survey of some of the major concerns in this field are treated below, at the beginning of the third section.

The Contents of the Book

This book treats the central issues facing mathematics teaching today, and indeed tomorrow. The three sections of the book, and the issues treated in them, are as follows.

Part I. Innovations in the teaching of mathematics: an overview; the new technologies, including the microcomputer and interactive video; the new forms of assessment, including classroom-based assessment and investigations; and new curriculum developments in school mathematics, including the contextualized 'Enterprising Mathematics' Project.

Part II. New research perspectives on the learning and teaching of mathematics: an overview; detailed expert scrutiny of aspects of the primary and secondary mathematics curriculum; a critical examination of some of the sacred cows of mathematics teaching, including discussion and the use of concrete experiences; and finally, the constructivist view of the learning of mathematics applied to the teaching and assessment of mathematics understanding.

Part III. The social context of mathematics teaching: an overview; the key issue of gender and mathematics; the implicit social and political values of mathematics and mathematics education; multicultural and anti-racist aspects of mathematics teaching, and the mathematics teacher.

The teaching of mathematics has never been in such a state of ferment and change as it is now. Britain is facing major reorganization at all levels in education, mathematics teachers are to be regularly appraised, the mathematics curriculum is to become centrally directed and tested, and the shortage of qualified mathematics teachers continues to grow. At the same time as these pressures grow many exciting developments in the teaching of mathematics are occurring. The new information technologies, which offer mathematics teaching so much, continue to develop and permeate all levels of society. Better forms of teaching and assessment in mathematics continue to spread. The demand for in-service courses continues to grow. The mathematics teaching organizations such as the Association of Teachers of Mathematics, The Mathematical Association, the Institute of Mathematics and Its Applications and the Mathematics Instruction Sub-committee of the Royal Society increase their already considerable activities for the enhancement of the teaching of mathematics. New curriculum projects which may further transform the teaching of mathematics are underway. Research in mathematics education continues to deliver new knowledge and exciting new perspectives from which to view and understand mathematics classroom teaching. Mathematics education is becoming increasingly accepted as an important area of knowledge in its own right. Thus, perhaps because of all of the change, development and ferment, there never has been a more exciting time to be involved with the teaching of mathematics than the present — except, perhaps, the future....

References

Assessment of Performance Unit (1985) *A Review of Monitoring in Mathematics 1978–1982*, London, Department of Education and Science.

Bell, A.W., Costello, J. and Kuchemann, D. (1983) *A Review of Research in Mathematical Education, Part A, Research on Learning and Teaching*, Windsor, NFER-Nelson.

Booth, L. (1984) *Algebra: Children's Strategies and Errors*, Windsor, NFER-Nelson.

Brown, J.S. and Burton, R.B. (1978) 'Diagnostic Models for Procedural Bugs in Basic Mathematical Skills', *Cognitive Science*, 2, pp. 155–92.

Carpenter, T., Moser, J. and Romberg, T. (1982) *Addition and Subtraction: A Cognitive Perspective*, Hillsdale, N.J., Erlbaum.

Christiansen, B., Howson, A.G. and Otte, M. (1985) *Perspectives on Mathematics Education*, Dordrecht, Reidel.

Cockcroft, W.H. (1982) *Mathematics Counts*, London, HMSO.

Davis, R.B. (1984) *Learning Mathematics: A Cognitive Science Approach*, Beckenham, Croom Helm.

GOOD, T.L., GROUWS, D.A. and EBMEIER, H. (1983) *Active Mathematics Teaching*, New York, Longman.
HART, K.M. (1981) *Children's Understanding of Mathematics: 11–16*, London, John Murray.
HART, K.M. (1984) *Ratio: Children's Strategies and Errors*, Windsor, NFER-Nelson.
HER MAJESTY'S INSPECTORATE (1985) *Mathematics from 5 to 16*, London, HMSO.
HOWSON, A.G. (1982) *A History of Mathematics Education in England*, Cambridge, Cambridge University Press.
KERSLAKE, D. (1984) *Fractions: Children's Strategies and Errors*, Windsor, NFER-Nelson.
LOW ATTAINERS IN MATHEMATICS PROJECT (1987) *Better Mathematics: A Curriculum Development Study*, London, HMSO.
NATIONAL COUNCIL OF TEACHERS OF MATHEMATICS (1980) *An Agenda for Action*, Reston, Va., NCTM.
NATIONAL CURRICULUM MATHEMATICS WORKING GROUP (1987) *Interim Report*, London, Department of Education and Science.
SECONDARY EXAMINATIONS COUNCIL (1985) *The National Criteria for the General Certificate of Secondary Education*, London, HMSO.
STEFFE, L.P., von GLASERSFELD, E., RICHARDS, J. and COBB, P. (1983) *Children's Counting Types: Philosophy, Theory, and Application*, New York, Praeger.
TASK GROUP ON ASSESSMENT AND TESTING (1987) *A Report*, London, Department of Education and Science.

Part I

Innovations

In the introduction the immediate roots of the current British scene in mathematics teaching were found to lie in the 1970s. In relation to innovations in the teaching of mathematics a more appropriate starting point is the 1960s, for this was also a time of innovation in the teaching of mathematics. It was the decade of:

- the broadening of the mathematics curriculum to include new topics, both pure and applied;
- the inclusion of computer aware topics in mathematics, such as flow charts and base two numeration;
- the application of technology to the teaching and learning of mathematics via teaching machines and programmed learning;
- curriculum development projects in mathematics, including the Nuffield Project at the primary school level and the School Mathematics Project (SMP) at the secondary level;
- progressive teaching methods, including discovery learning, and the beginnings of interest in problem-solving and mathematical investigations;
- the progressivization of assessment, including the move to abandon the 11+ examination for entry to secondary schools and the introduction of the Certificate of Secondary Education in mathematics for pupils of average, or just above, attainment in mathematics (and other subjects).

Thus many of the present themes of innovation emerged during the 1960s: new curricula, new technology, new teaching methods and new systems of assessment. Curriculum commentators have pointed out that this pattern can be described as pendulum-like, the swing from one extreme position to the opposite. The progressivism of the 1960s was followed by a reaction in the 1970s — the back-to-basics movement — which in turn was followed by a new wave of progressivism in the 1980s. Does this mean that there will be a further reaction in the 1990s against what are seen as the progressive excesses of the 1980s? Whilst such a reaction is a possibility, there are several new factors which suggest that we are not trapped in one of Vico's cycles, doomed to replay history.

> There is a new realism. Those involved in education accept their accountability to parents and authorities as a matter of fact.
>
> The 'goods' of education (i.e., the desired outcomes) have been redefined to include more than the basics. Employers and authorities agree that in addition to literacy and numeracy (and graphicacy and computeracy) school leavers need presentation, communication and decision-making skills, and the ability to solve problems, to participate in teamwork and to work cooperatively. These are the skills addressed by contemporary mathematics teaching at its best.
>
> The mathematics teacher has become increasingly professionalized with an all-graduate entry to the profession. Teacher education requires four years of advanced study for an honours degree in education, or a three-year BA or BSc followed by one year of study for a Postgraduate Certificate of Education (PGCE).
>
> Mathematics education has also become more professionalized, as the growth of mathematics teaching and research journals since the 1960s indicates. The volume and standards of research on the teaching of mathematics have both risen, and practitioners are better informed of its results and of the justification for contemporary practice.

Provided the mathematics teaching community sustains and enlarges these aspects of accountability, justification and professionalism, there should be no reason for the pendulum to swing against current innovations.

Some of the changes are in keeping with broader social changes. Citizens at large seem more inclined to question the arbitrary dictates of authority. This is reflected in the classroom, where teachers no longer have the unquestioning obedience of their pupils. Thus to sustain an authoritarian model of mathematics teaching in the school of today would be increasingly difficult. As was said in the introduction, one of the key innovations in the teaching of mathematics has been the liberalization of teaching methods. The new technology moves the focus of attention away from the teacher to the electronic medium, and encourages the role of teacher as facilitator instead of all-seeing, all-powerful authority. New curricula and new assessments also encourage a variety of teaching styles including co-inquirer, facilitator, manager and explainer, as well as the traditional instructor role. For these innovations explicitly require higher level responses on the part of learners, responses which cannot be produced in response to the commands of an authority figure.

The innovations currently underway in the teaching of mathematics are new technology, new curricula, new assessments and, pervading all of these, new teaching approaches. These mean that mathematics teaching in the 1990s will be substantially different from how it has been in the 1960s and 1970s, and even in the 1980s. These innovations augur well for the teaching of mathematics. The new technologies and curricula will add excitement and relevance to the teaching of mathematics. The new modes of assessment, particularly the GCSE examination, are more searching in their demands on learners. The broader range of teaching styles provides a much better opportunity to achieve the goal of preparing confident problem-solvers who are capable of independent critical thought. Surely this augurs well not only for the outcomes of mathematics teaching, but also for society as a whole.

NEW TECHNOLOGY

The most important development of the 1980s for the teaching of mathematics has been the advance and spread of the products of the new technology. These include the electronic calculator, the microcomputer and interactive video systems, as well as video recorders, programmable robots such as the Turtle, and other devices. The impact of these products on the mathematics curriculum can be felt in two ways: on the content and on the modes of learning and teaching.

First of all, there is the impact on the content of the mathematics curriculum. The universal adoption of new technological products, especially the electronic calculator and the computer, means that knowledge, familiarity and competence in using these resources is a required part of modern education — especially in mathematics. The functions of these resources also render much of the traditional mathematics curriculum obsolete. With electronic calculators (and computer spreadsheets), whole number, fractional and decimal computation need not be carried out laboriously by hand. The years devoted to acquiring the arithmetical algorithms involved in long-hand computation can no longer be justified. Electronic calculators can also carry out algebraic manipulation, and readily available computer software renders the manipulative skills of algebra, trigonometry and calculus unnecessary. Thus a substantial part of the mathematics curriculum — that concerned with mechanical computation and manipulation — is effectively obsolete. Yet as we enter the 1990s this fact of modern life is not widely recognized.

The new technology also requires a new curriculum emphasis if it is to be used effectively. Learners need to be able to interpret and check numerical results, tables and graphs, they need to be able to think procedurally, to write and debug programmes, and above all else to be confident and creative problem-solvers.

The second way that the impact of the new electronic technology is felt is on the modes of teaching and learning mathematics. With the calculator or computer in the classroom, the teacher is no longer the sole arbiter of truth. Although they can be used in a variety of ways, the new electronic resources encourage an exploratory approach to the learning of mathematics. The calculator, microcomputer and interactive video systems all invite exploration by the learner. Also the better examples of software written for microcomputers and interactive video are designed to promote creative, higher level thought. Computer programming — whether in BASIC, LOGO, Prolog, APL or whatever language — is of itself a high level problem-solving activity.

The introduction to this part of the book warned of the dangers of the pendulum swinging back, away from the progressive innovations presently facing the teaching of mathematics. The one area which has suffered no swing of the pendulum is the area of technology. Parents, employers and teachers alike have recognized the crucial role of the new technology in society. They all agree that school is the place, and the mathematics classroom in particular, where children need to be prepared for the world of tomorrow, especially its technological aspects.

The three chapters deal with aspects of the role of the new technology in the teaching of mathematics. They treat the role of the microcomputer in primary mathematics, the utility of the language LOGO, and the most recent of the technological resources, interactive video. The chapters raise many of the questions concerning the potential of the new technology for the teaching of mathematics, and provide answers to some of the questions. Needless to say, many areas are left untouched; the role of the electronic calculator, computer software, such as spreadsheets, databases, further computer languages, the potential of electronic communications media and a host of other areas. However, the chapters that are included begin to show some of the potential of the new technology for the teaching of mathematics that is currently being realized.

1 *The Role of the Microcomputer in Primary Mathematics*

Paul Ernest

The last decade has seen the spread of a new resource with the power radically to transform the teaching and learning of mathematics in primary schools: the microcomputer. Although the Cockcroft Report (1982) on the teaching of mathematics devoted a chapter to computers and calculators, there was no widespread adoption of the microcomputer until the Department of Trade and Industry initiative put one in every primary school. By 1985, when *Mathematics from 5 to 16* (HMI, 1985) was published, the list of objectives for mathematics education from five years onwards included the skill:

> Objective 9. the use of a microcomputer in mathematical activities

and the criteria for choosing the content of the mathematics curriculum included:

> Criterion 10. The content should be influenced increasingly by developments in microcomputing.

The second edition of this document incorporates the responses of the education community, indicating a broad measure of support, and concludes by recommending that the teaching 'profession needs to maintain its efforts to ... develop the purposeful use of ... microcomputers to transform mathematics teaching and learning' (HMI, 1987, pp. 14–15). The Mathematics Working Group (1987) *Interim Report*, on which the *National Curriculum in Mathematics* is to be based, refers to the role of microcomputers on five of its twenty-two pages, and states: 'the provision of the national mathematics curriculum should ensure ... the use of ... microcomputers as appropriate' (p. 5); 'It is the view of the working group that mathematics is learned by ... using various resources ... [including] ... microcomputers' (pp. 6–7). Thus teachers of mathematics at all levels, including primary school teachers, can expect to see a greater prominence given to the use of microcomputers in the teaching and learning of mathematics. The same is likely to be true across the whole of the *national curriculum* for primary schools, but this chapter will deal only with mathematics.

The pressure to include more use of the microcomputer in primary mathematics raises a number of questions: What does the microcomputer offer the teaching and learning of mathematics in the primary school? How can it best be used? Where can suitable resources be found, especially software? This chapter tries to answer these questions. To illustrate the uses of the microcomputer a number of specific examples of software are referred to. These are mainly drawn from the MEP pack, *Primary Maths and Micros* (see Appendix). But first we consider the background to these innovations.

The Impact of Electronics on Primary Age Children

Electronic resources — modern microchip and information technologies — are impinging on school children in a number of ways. Although the focus of this chapter is the microcomputer, the range of new resources is far wider. Children are meeting digital displays and graphical displays, and programming many electronic devices, including the following: digital watches, video recorders, electric cookers, microwave ovens, microchip toasters, petrol pumps, electronic calculators, electronic weighing machines, electronic checkout tills, hi-fi systems, telephones, hand-held video games, video arcade games, electronic teaching aids such as 'Speak and Maths' and 'Little Professor', programmable toys such as 'Big Trak', microcomputers, word processors, interactive video systems (such as the Domesday disc), microcomputer controlled robotics. This list, although far from complete, begins to show the impact that the electronic revolution is having on the experience of primary school children. We can expect this impact to accelerate as the children of today approach adulthood in the twenty-first century. Already this impact calls into question some aspects of the mathematics curriculum, as *Mathematics from 5 to 16* (HMI, 1985) anticipated.

> How useful is telling the time on a traditional circular display? Digital displays of time now dominate the child's experience.
>
> How useful is proficiency in operations on fractions? Decimal fractions wholly dominate commerce, industry and higher education.
>
> How necessary are the traditional computational algorithms (the four rules)? Calculations needed outside school are done mentally or with an electronic calculator, or with a computer spreadsheet.

> How well are children prepared for the computer age? Are they being taught to read information in graphs and tables? Are they learning to think procedurally, in terms of sequences of operations? Are they acquiring a feel for large numbers and for variables? Are they learning to formulate and solve problems, and to check their solutions for reasonableness?

This chapter will not answer these questions (but for the beginnings of an answer see Shuard, 1986). However, one thing is certain: children will need to be familiar with the products of the electronic revolution as preparation for living and working in the world of the twenty-first century. Foremost of these products is the microcomputer, which has unlimited potential to change the worlds of commerce, industry, government and education. As indicated, we will focus on only one part of this last area: the potential of the microcomputer for the teaching and learning of primary mathematics.

The Microcomputer in Primary Mathematics

The microcomputer has the power radically to transform the teaching of mathematics in the primary school, because it offers both teacher and learner new ways to approach mathematics. Some of these new ways, such as the introduction of more problem-solving, investigation and mathematical discussion, have been recommended independently of the microcomputer. The microcomputer is a powerful resource for facilitating and implementing these approaches beyond normal classroom practice. In other ways the microcomputer offers the learner genuinely novel experiences: access to controllable graphics and animation, interaction with an almost infinitely variable medium, instantaneous feedback and the non-judgmental correction of mistakes, mathematical adventure gaming, the opportunity to program the computer in languages such as LOGO and Basic, to name but a few of the experiences available. These teaching and learning approaches are now available to primary teachers with very little expertise and experience. More important are curiosity — a willingness to explore the medium — and humility — the acceptance that our students will rapidly outstrip us in knowledge and expertise.

We begin to consider the approaches that the microcomputer makes possible by classifying the different ways in which it can be used in the classroom. First of all, it can be controlled by the *teacher* or by the *learner*. The distinction is between the microcomputer being used by the teacher for demonstration and discussion purposes, or the children using it for themselves in their learning (interacting 'hands on', as it is termed). A second distinction can be drawn between two modes of use: the difference between *using software* — using ready made programs and soft-

Table 1. Different Types of Use of the Microcomputer in Mathematics Teaching

Control / *Mode of use*	*Teacher* (*Teacher demonstration*)	*Learner* (*Children 'hands on'*)
Using software	Demonstration Simulation Posing-problem Starting investigation Data display	Concept learning Problem-solving Exploring, investigating Game playing Adventure gaming Data display Skills practice
Programming the computer	Short programs Calculating on screen Graphics	LOGO programming Basic programming Short programs Projects

ware — and *programming* the computer. These two sets of distinctions are combined in Table I.

This table shows a sample of the more common uses of the microcomputer in primary mathematics. It is far from complete, and enthusiasts will know that there are very fruitful activities involving the creation and use of databases, spreadsheets, robotics, Prolog programming, and so on, beyond what is mentioned, which will not be discussed here. This still leaves the vast range of activities under the headings listed in the table.

The view adopted here is that each of the four categories shown — teacher-led use of software, teacher programming, child-led use of software, child programming — has a great deal to offer. Unlike Seymour Papert (1980), who argues that using software is 'letting the computer program the child', the present author's view is that good software can be used very fruitfully in the primary classroom. In fact, given that most primary school teachers will not initially have the interest, confidence or skills to program the microcomputer, the uses of software are likely to provide the most widespread and effective approaches to microcomputer use. Certainly they represent the more accessible ways of introducing the microcomputer into primary mathematics. We consider each of the four main categories in turn, but for the above reasons focus mainly on the uses of software.

Teacher-led Use of Software

There are a number of ways that a teacher can initiate class activity through the use of software, including: demonstration, simulation, problem-posing, starting an investigation, and data collection and display. In addition, the teacher can bring the class together during topic or project work, for example, and use the microcomputer to summarize and display the data generated by the children. We examine a number of these specific types of use.

Demonstration

The teacher can introduce children to an area of activity by means of a demonstration of some computer animated graphics, followed by a class discussion. An example is provided by the MEP program, 'HALVING', which shows the attractive and increasingly complex division of a square into a pattern of blue and red regions (of equal areas). The program can be stopped at any point (by pressing space bar) for closer examination and discussion, and the children asked, for example:

Figure 1. HALVING: A Concept Web

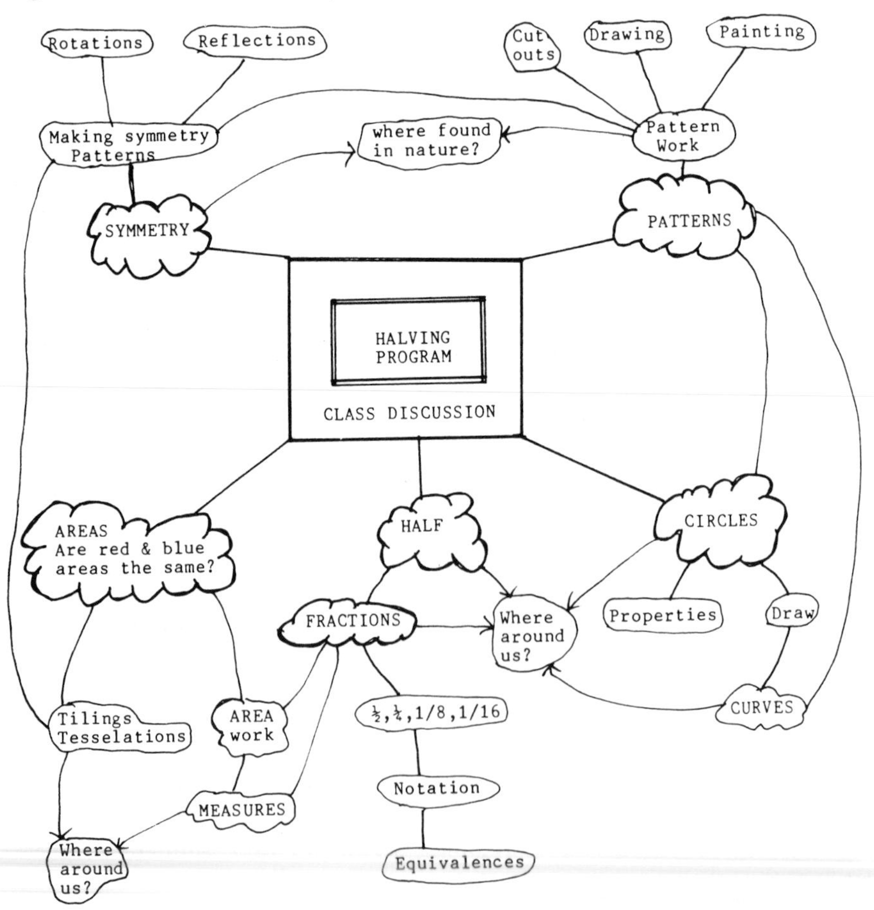

What pattern is showing? What shapes?
What mathematical ideas does it illustrate?
What fraction of the whole square is the red region?
Are you sure?
What argument could you use to convince somebody else of this?

'HALVING' provides a stimulus for a whole class discussion which could include halves, fractions, patterns, circles, symmetry, tilings, areas, shapes, and so on. Figure 1 shows a concept web of how these ideas can be developed from a class discussion of 'HALVING'. It has been successfully used as an introduction to project work, with children making their own patterns, and to a discussion of how the ideas generated relate to the children's experience and environment.

The use of software for teacher demonstration is most likely to be effective if there is a full discussion, and then the children engage in related follow-up work, as is suggested in this example. Simulation software (which involves sequences of images simulating some real world events) can be used similarly.

Problem-posing and Starting an Investigation

The teacher can use software to pose a problem or to start an investigation. For example, the MEP programs, 'PATTERNS 2' and 'FIND ME' pose problems. In 'PATTERNS 2' the teacher, or children, choose the size of a large equilateral triangle grid. Then they have to guess or predict the number of triangles in the bottom row. Thus if a triangle size seven is chosen, a display like that in Figure 2 is shown. The teacher can type in children's suggested answers to as many different sized triangles as is desired. Thus in the case of the illustration the number of triangles in the bottom row is thirteen. Then the teacher can set the children the problem of trying to find a rule for predicting the number of triangles in the bottom row. They can try out further examples on isometric paper (triangular grid spotted paper). When they are ready and think they have found the rule, the teacher can test suggestions by means of an option in the program.

The program also offers a second problem which can be approached similarly, involving the number of small triangles covering the whole of the triangular grid. (In the case of the illustrated figure, there are forty-nine altogether). Naturally, this is only one way of using the software. Variations enjoyed by children include having one child (or several) key in the suggestions, or carrying out the whole exercise as a class question and answer activity.

The potential of this program, let alone this type of software, may not be clear from this brief account. What the program offers specifically is an approachable but challenging problem which provides the opportunity to utilize the following processes and strategies:

Figure 2. A triangular grid size 7 (display in PATTERNS 2).

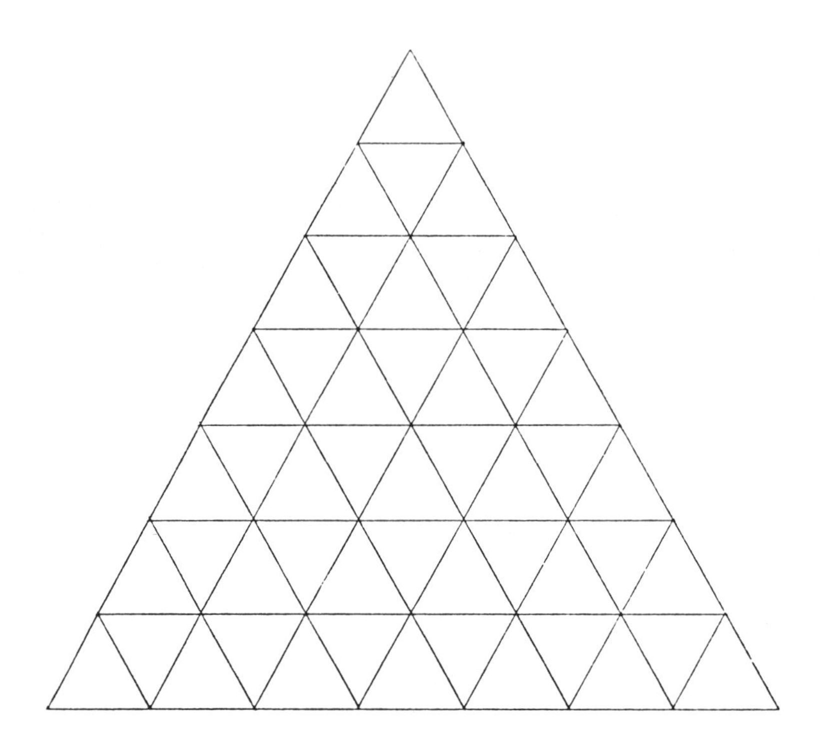

visualization skills and visual thinking in solving the problem;
trial-and-improve strategies;
making conjectures;
testing conjectures;
generalization from numerical and visual patterns;
applying generalizations to specific examples;
formulating and explicitly stating relationships;
for the more advanced, stating relationships abstractly in the language of algebra (in the example the rule is 2N-1.)

As well as developing strategies such as these, problem-solving software poses problems which can serve as starting points for investigations if, after discussion, children are encouraged to extend the ideas in them in their own directions. Note that the programs mentioned above are included in the MEP Primary Maths and Micros Pack (see Appendix), which also contains suggestions for follow-up activities.

Displaying Statistical Data

A further use of the microcomputer in the mathematics classroom is for the display of statistical data in the form of graphs. A simple program of this type is 'DATASHOW'. This allows numerical data (in up to eight categories) to be shown in the following forms: a table of results, a bar chart (horizontal bars), a pie chart, and to be sorted by numerical size or alphabetically. Data can be put in to show how many children have cats, dogs and other pets; or how far eight children jump in a long jump contest; or how many hours a child spends sleeping, watching TV, at school, and in other activities; and so on. Displaying sets of data like these is interesting and teaches ideas of graphical representation. The facilities available in this program mean that data can be rapidly transferred from one mode of display to another, so that the class can discuss which mode best suits that set of data. All too often discussion of the meaning and the most useful ways of displaying data is treated too lightly, if at all. Software like 'DATASHOW' facilitates class discussion of these issues. This can be followed up by a discussion of examples of the tables, graphs, facts and figures used in advertizing and in the media in general. If children are to become full participants in our democratic society, they need to be able to understand, discuss and critically evaluate the claims made in the media, supported with statistical displays. Software like 'DATASHOW' provides an introduction to statistical graphs, and allows children access to pie charts, which can involve very complex calculations otherwise.

'DATASHOW' is another of the programs from the MEP Primary Mathematics and Micros Pack, like the programs discussed above. Although fairly simple, it offers quite a lot. There are, of course, far more sophisticated database programs for primary schools (e.g., 'GRASS' and 'QUEST'), which not only allow many more data to be represented, but permit the elaborate classification and sorting of the data as well.

These three examples give an indication of the uses to which the teacher can put software in a whole class setting. The shared feature is that teacher demonstration — for want of a better word — can provide a shared stimulus for class discussion and activity. This serves to counterbalance the common practice of letting children learn their mathematics individually from published schemes, such as SMP 7–13, Fletcher, Ginn, Peak, Nuffield, SPMG, and so on. As the Cockcroft Report (1982, paragraph 243) strongly recommends, in addition to the exposition, reinforcement and practice that the schemes provide, children at all levels need to engage in practical work, problem-solving, investigational work and above all else discussion in their learning of mathematics. Teacher-led microcomputer demonstrations, leading to discussion, provide one accessible way of redressing this balance.

Learner-led Use of Software

Most of the classroom uses of mathematics software treated in the discussion of teacher-led computer work are also available to children when they are working directly with, or controlling the microcomputer themselves. Programs can pose problems and provide the starting points for investigations, perhaps one of the most fruitful of all the uses. The micro can provide simulations for children to watch, interact with, and discuss in groups. Children can use statistical data display programs, such as 'DATASHOW', to show the results of data gathering project work. These types of use include some of the most powerful learning experiences we can give children. But they have each been discussed and exemplified above.

In addition to these, there are other worthwhile types of use, considered below. In each case a whole class of children could work at software if a computer laboratory were available. While this can be arranged by taking the class on a visit to a local secondary school or college, the reality of the primary school is usually different. Most primary teachers have access to a single microcomputer in school (often for only one week per half term). But this can still be used to give all children in a class a chance for a hands-on experience. What is often arranged is a rota, whereby small groups of children take turns working with the microcomputer, discussing what they are doing together. Over a period of time this can add up to a substantial experience for each child.

Concept Learning

Some software helps children to acquire or develop their understanding of mathematical concepts. Two examples of such software are 'SUBGAME' and 'SIZEGAME', both of which help children to develop their understanding of the crucial concept of place value in number (numeration, to be more precise). 'SIZEGAME' employs a game format, and requires children to order random digits to achieve the largest number. This leads to, or reinforces, the notion that value is not just a question of digit size, but also of placing. 'SUBGAME' also involves random digit placing in a column subtraction to achieve the greatest difference, again teaching place value. The concept of probability is also developed by these programs, because random digits are generated, and the player has to make decisions, such as whether the occurrence of large or small digits is probable. Thus these programs help to develop the concepts of place value and probability in a motivational setting. Beyond this the advantages of software like these two programs include the instantly self-checking feature, as well as the fact that they encourage discussion and cooperation among children.

This feature of self-checking is one of the great strengths of concept learning with the microcomputer, for it means that a child's application of a concept is immediately reinforced and guided along the intended paths. The computer is also infinitely patient, allowing the child to take his or her own time at a certain level of working. Finally the computer is non-judgmental: children do not feel the same stigma in making mistakes with a computer. There is no record of failure in their book, no series of written crosses, no sense of letting another person down, no sense of failure. All of these feelings in the normal classroom are what can contribute to many persons' feeling of failure and inability in mathematics. Working with a microcomputer can help to stave off these very negative reactions felt by some.

Game Playing

We have already seen that concept learning software can be in game format, in the case of 'SUBGAME' and 'SIZEGAME'. These allow different numbers of children to play together. Another example is 'TEASHOP', which could also be described as a simulation (since it requires the player to plan a certain number of teas for a school cricket match, and sales depend on how the weather turns out, which the computer decides). This game can be played by a single child or by several. It reinforces arithmetic skills and calculations involving money, as well as strategic thinking, and even basic ideas of probability. Since the program requires children to try to maximize their business profits, can it be said to encourage entrepreneurship and the spirit of the enterprise culture? A successful strategy involves maximizing both the cost of a cup of tea as well as the number prepared for sale, provided the weather outlook is good. This could be used to raise the question of the morality of maximizing consumer costs to increase business profits. Or is this too contentious an approach for the primary classroom? While this issue can be debated by teachers, children find the game most motivating and involving.

Adventure Gaming

Adventure games both in book and software form are increasingly popular with children, many of whom love to immerse themselves in the dream world that the games involve, and to master it. A simple adventure game is 'MALLORY', which resembles the commercial game, 'Cluedo'. Although little of the content is explicitly mathematical, the thought processes involved in using it are problem-solving strategies such as using trial-and-improve, looking for pattern, logical thought, elimination of possibilities, making conjectures, testing them, and so on. Other adventure games like 'L — A Mathemagical Adventure' help to develop spatial skills (using the points of the compass, mapping the imaginary land), reading skills (most information is presented in writing) and many other incidental mathematical ideas. The key strength of adventure game programs is the motivation they engender in children. They also provide a link between mathematics and the skills of reading and writing. A natural extension activity to adventure gaming is for children to write a continuation of the story. However, because of their complexity they may cause difficulties with the young or low attaining child.

Skills Practice

Many of the programs discussed above involve the practice of skills, especially basic numerical skills. This is true of 'SIZEGAME', 'TEA-SHOP' and others. Another skills practice program is 'PLAYTRAIN', in which children have to distribute a number of passengers among the carriages of a train uniformly, or obeying some other numerical constraints. The program explicitly requires and develops basic number skills, but these are mental as much as written arithmetical skills, and the whole program could also be classed as problem-solving. In the view of the author there is no need for skills practice to be just a computerized page of 'sums'. This may ultimately take away children's motivation, as well as reinforcing a view of mathematics as a collection of pedestrian and routine activities. It is also a wasted opportunity to develop children's higher level thinking, which as well as the acquisition of knowledge must be the goal of teaching mathematics.

It should be clear that software can be classified in a number of ways, according to many factors including the type of learning activity (such as skills practice, concept learning, problem-solving and investigating), the software format (e.g., game playing or adventure gaming), the topic involved (data display), and there are many other factors such as age suitability, quality of the software and ease of accessing the ideas involved, which have not been referred to. This chapter has not tried to provide a taxonomy of software types, but has introduced some of the more important distinctions between, and uses of, mathematics software in the primary classroom. These have been illustrated with specific examples that can be tried out with children.

Programming the Computer

In addition to teacher- and child-led use of software, Table 1 shows a range of programming activities which can enhance the teaching of mathematics in the primary school. The teacher can calculate directly on the screen, write short programs in the computer languages Basic or LOGO or even Prolog, and so on. The children may be able to do the same, and there is some very exciting work going on in parts of Britain with quite young (infant school) children programming in LOGO. Children's use of LOGO may be best started with the teacher structuring some preliminary (non-computer) activities, and then demonstrating some of the basic commands. However, the key element of the LOGO philosophy is that the children explore and use the language themselves. Some of what LOGO programming offers the teaching and learning of mathematics is treated in Ernest (1988). The other forms of teacher and child programming listed will not be dealt with here; the enthusiast can find these areas treated well elsewhere. A final word on programming: these activities may unleash the full potential of the microcomputer as a learning aid in mathematics, and across the whole primary school curriculum. Currently there are signs that the governments, through its various agencies, will be pressing for much more emphasis on information technology, especially the programming of computers, in initial teacher training. Thus it may come to pass that all new entrants to the profession brings these skills with them into schools. Such a development may radically alter the face of primary school teaching, as visionaries such as Papert (1980) foretell. Meanwhile, the use of software offers both teachers and learners a very valuable learning experience, and serves to introduce both groups to some of the potential of the microcomputer.

Benefits of the Microcomputer

Some indication of what the microcomputer offers the teaching and learning of primary mathematics, in terms of the range of uses, has been given above. Beyond this the microcomputer has tremendous classroom potential because it is:

motivating — children find microcomputers fascinating;
varied — the range of worthwhile learning experiences now available on the BBC B+ microcomputer, for example, is vast;
flexible — good software (or programming languages) allows children to enter at the level that is most suitable and comfortable for them;
interactive — it offers an almost infinitely variable learning experience within one situation or set of situations, according to the child's responses;
self-checking — children's mistakes are diagnosed immediately, and they have the opportunity to remedy them;
non-judgmental — children feel no stigma in making mistakes with a computer — it is infinitely patient and non-condemnatory;
visual — it offers exciting displays which aid spatial visualization abilities;

powerful — it offers new learning experiences that have simply not been available before;

learner-centred — the micro can support a shift towards a learner-centred approach, giving children more control over their learning of mathematics.

For all these advantages, teachers unfamiliar with the microcomputer may be reticent to try it out in the classroom, because of a lack of experience and skills. Teachers need to be made aware of the fact that every classroom is already equipped with a readily available computer expert, in amongst the children. Every user of computers in education knows that no matter how much one knows, it is only a matter of time before at least some of the children become more proficient at using microcomputers. But then that is education at its best, is it not?

Conclusion

This chapter has looked at a key electronic resource for teaching mathematics, the microcomputer, which has revolutionized practice in the world outside school, and will continue to do so increasingly as we approach the twenty-first century. The microcomputer is becoming so important in modern life that 'Computer Literacy' — as familiarity and confidence with computers are called — has been classified as a basic skill which all children need to acquire.

Some indication of its immense potential for the teaching of mathematics has been given, including its possible benefits. However, the use of these resources needs to be built into our teaching programmes, so that their benefits are utilized and not marginalized. As the reports quoted in the introduction suggest, we also need to reflect on the mathematics curriculum we are providing children in the light of modern developments and thinking. Are we empowering them to be the confident problem-solvers that fulfilling themselves and society requires? Will they be confident wielders of the new technology?

References

Cockcroft, W.H. (1982) *Mathematics Counts*, London, HMSO.

Ernest, P. (1988) 'What's the Use of LOGO?', *Mathematics in School*, January 1988, pp. 16–20. (Chapter 3 of this book.)

Her Majesty's Inspectorate (1985) *Mathematics from 5 to 16: Curriculum Matters 3*, London, HMSO.

Her Majesty's Inspectorate (1987) *Mathematics from 5 to 16: The Response to Curriculum Matters 3*, London, HMSO.

National Curriculum Mathematics Working Group (1987) *Interim Report*, London, Department of Education and Science.

PAPERT, S. (1980) *Mindstorms*, Brighton, Harvester Press.
SHUARD, H. (1986) *Primary Mathematics Today and Tomorrow*, London, Longman.

Appendix: Sources of Good Mathematics Software

MEP (Microelectronics in Education Programme):
MEP *Primary Maths and Micros*
MEP *Infant Mathematics Software*
MEP *LOGO Inservice Pack*
MEP *Problem Solving with a Micro*
The first of these packs contains most of the software referred to above. The packs have been issued to every LEA.

SMILE Group *The First 31 Programs*, and *The Next 17*, London, ILEA Learning Resources Unit.

ASSOCIATION OF TEACHERS OF MATHEMATICS *SLIMWAM 1 and 2 (Some Lessons in Mathematics with a Microcomputer*), Derby, ATM.

STRAKER, A. *Mathematical Games and Activities* (*Volume 1*), and *Mathematical Investigations in the Classroom*, Anita Straker-Capital Media.

ASSOCIATION OF TEACHERS OF MATHEMATICS *L—A Mathemagical Adventure*, Derby, ATM.

STRAKER, A. *Martello Tower* (in the MEP Problem Solving Pack).

Longman Software *The Kingdom of Helior*, London, Longman.
The last three are examples of mathematical adventure game software.

2 *Using Interactive Video in Secondary Mathematics*

Dudley Kennett

For many years now computer software has been developed for use in education and training. We have recognized the strengths of this technology and looked for ways of overcoming its weaknesses. A particular weakness has been the quality and control of visual displays and sounds. Laser technology now enables us to move forward. It gives us the opportunity to have moving video images, computer text and graphics, all under the control of a low cost microcomputer. Central to this development is the videodisc, each side of which can hold up to 54,000 frames or thirty-six minutes of sound tracks. This can all be controlled by computer, has fast random access and very high quality of sound and vision. It is known as Interactive Video (IV).

A typical workstation is illustrated in Figure 1. It is made up of a videodisc player, monitor, microcomputer, disc drives and an overlay box. The overlay box combines the video outputs of the microcomputer and videodisc player for display on the monitor. A complete system can cost from £5000 to as little as £3000, and there are clear signs that £2000 will be a typical price in the near future. At such prices users in education may be interested in using IV, but they need to have material to use on these systems. Since 1982 and until recently software for IV has been developed almost exclusively for industry and commerce. Training departments have found it to be an effective and cost efficient method of training, and in marketing it is used for point of sale and other promotions. It is quite clear that managers and accountants believe that this is worthwhile. They are investing in software and hardware, and expect a return on their money. In education there is a problem. Investment in the development of curriculum materials is usually on a small scale, and our objectives are less clear than those of the trainer. One step to help overcome this was taken in 1985 when the DTI set up the Interactive Video in Schools Project (IVIS).

Figure 1. The Layout of an Interactive Video Workstation

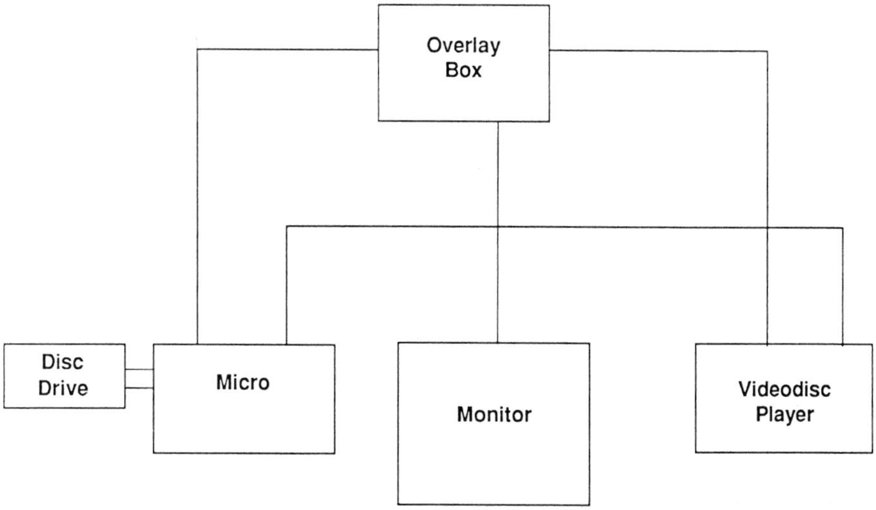

IVIS is a research and development project which aims to answer the question, 'How can IV be used in the classroom?' It consists of a central coordinating team, eight project teams each producing and trialling an IV package, and an evaluation team. The eight project teams and their subject areas are:

Mathematics: University of Exeter
Environmental studies: Moray House College of Education
Modern languages: Shropshire LEA
Geography: Loughborough University
Design: Leicestershire LEA
Primary science: Bulmershe College
Teacher training: Bishop Grosseteste College
Social skills: Centre for Learning Resources, NI

The evaluation is being carried out by the Centre for Applied Research in Education, University of East Anglia, and a research report was published in April 1988.

As part of IVIS the Centre for Innovation in Mathematics Teaching at the School of Education, University of Exeter, working with Blackrod Interactive Services Ltd, has developed an interactive video package for use in the secondary mathematics classroom. The title of the package is 'School Disco'.

'School Disco' brings the mathematics of the real world into the classroom by placing students in an environment where they can see how the decision-making processes of a straightforward business situation require some basic mathematics for their efficient solution. When asked to plan and organize a school disco, they are faced with the problems of deciding where to hold the event, who to hire to play the music, what refreshments to provide, the price of the tickets and what publicity is needed. The consequences of their decisions are then seen when they view the success, or otherwise, of the event they have planned.

The curriculum content of the package includes basic arithmetic, statistics, graphs, modelling and optimization. It is aimed at students of average ability in the 14–16 age range. On starting the system, pupils are asked to type in the date and their names; after this they see an opening sequence which sets the scene. A headteacher advances them a float with which they start their enterprise, and from there they are moved to the 'Main Map', a diagram (Figure 2) which gives an overview of the system and from which they can choose the activity they wish to do first. There is no prescribed strategy, and the system includes many random events. The values of the variables change, and it is unlikely that anyone using it more than once will be faced with the same combination of situations.

As they move from one activity to another, the pupils have to collect, record and summarize information. Sometimes this involves straightforward arithmetic and recording in tabular form, on other occasions graphs have to be drawn and interpreted. Help is always available, and is in three forms: help with the system, help with the game, help with the mathematics. The last is presented automatically if needed. There are also guides to the additional activities, or follow-up materials, which include supplementary problems and investigations.

The design of 'School Disco' enables a teacher to use it in several different ways. At one level it can be presented to pupils as a game to be played fairly rapidly making naive decisions, but nevertheless reaching a satisfactory end point. This can be carried out by individuals or small groups of pupils. At another level a teacher might use it as the basis for a class lesson in which a single decision point is studied in detail, for example, the choice of ticket price for the disco. This decision is critical to the success of the venture, and the mathematics involved includes the consideration of a price-demand graph, which then leads to a price-revenue graph, and from there to the idea of a maximum possible value for the revenue for a given choice of venue and music. After several such sessions, looking at various decision points, the activity can again be based on individual or small group work, but this time the emphasis would be on making optimum decisions using the available mathematics.

We can attempt to classify IV packages by drawing upon the typology of computer-based learning activities. Much of the IV in training corresponds to computer aided instruction, in which the system transmits information and tests the learning. Sometimes this is seen as drill and practice, and sometimes it is considered as an intelligent programmed

Figure 2. Main Map

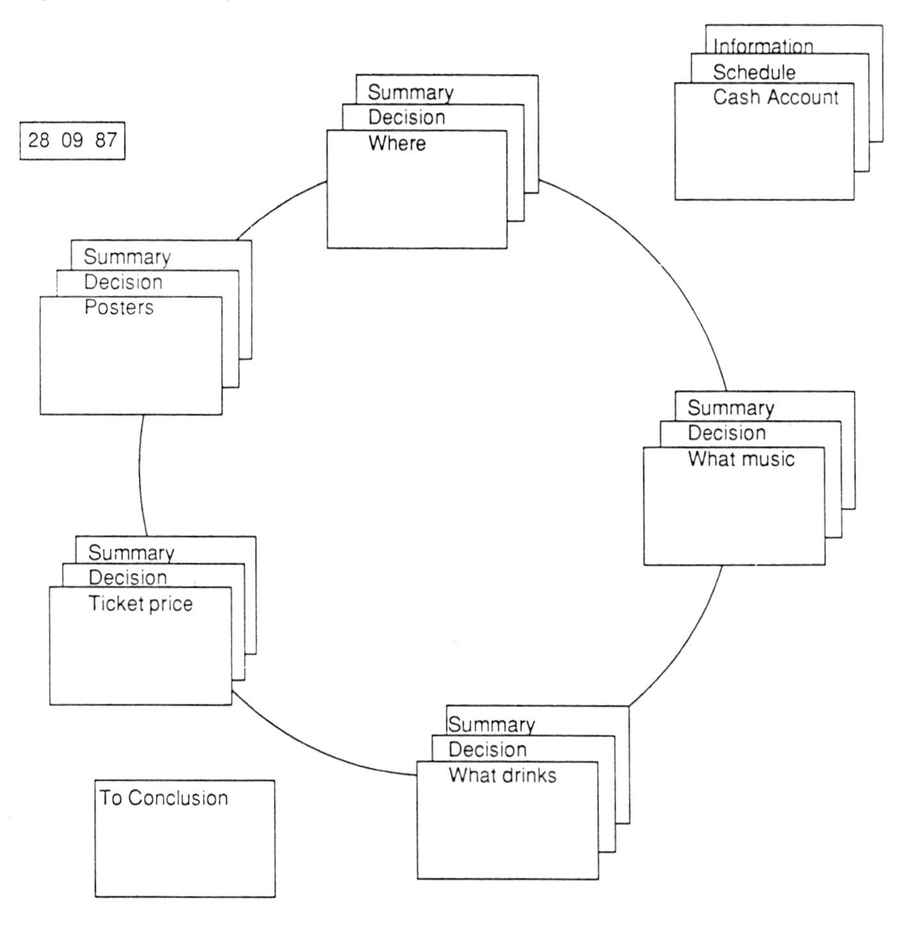

learning machine. Some would argue that as exposition is still the main teaching style for large parts of mathematics courses, and the rule-example-practice paradigm is frequently used throughout the mathematics classroom, then IV has a role to play. It could 'do' certain parts of the course. In the United States a commercial company, Systems Impact Inc., is marketing such material in a series called 'Core Concepts in Action'. A quotation from their sales literature gives an impression of the approach.

> Core concepts incorporates the principles of effective instruction; models proper techniques for presentation; focuses on important 'building block' concepts; uses motivating graphics which appeal to all age levels; and offers extensive 'branching' capabilities for remediation.

The series includes the titles 'Mastering Fractions', 'Mastering Decimals and Percents', 'Mastering Ratios' and 'An Introduction to Algebra'. This approach of exposition is used in parts of School Disco; for example, the Help modules are straightforward descriptions of the technique required for the current problem.

Computer-based simulations are a second category of computer aided learning. In these the learner can explore a model of a system while the computer acts as an umpire in a decision-making game. School Disco fits clearly into this category. The same can be said of the work on secondary mathematics of the Interactive Learning Group at the University of Newcastle. They too have chosen a game or simulation for their package, which develops the ideas of chance and probability for secondary school pupils.

A third category is software which provides an aid to conjectural learning. The best-known example is the computer language, LOGO. Using LOGO, a learner is supplied with an environment in which ideas and concepts can be explored. As yet there is no IV equivalent, but it is a prospect we can look forward to. Soon it will be possible to use the high quality images and sounds of IV to create a new generation of mathematical microworlds. The closest point reached so far in this development is the 'generic' type of package, in which the user is provided with a videodisc and software consisting of a wide range of images stored as a database. These can then be explored in an open manner, with suitable software in the form of an authoring language used to produce sequences of scenes and images as required. One example of this type is the IVIS Geography disc by Loughborough University. At Exeter we are exploring the possibility of producing similarly structured materials specifically for the use of mathematics teachers.

As yet we have no clear idea of the full impact of IV in the classroom. The IVIS trialling period is just starting. In education many have high hopes, but others may be rejecting the technology already. When looking at a package in isolation, it is important to remember that it is effectiveness in the classroom that is important. It is easy to be swept along by the excitement of the images, or be impressed by the technology, when it should be the outcome in terms of student learning that really counts. Alternatively, one aspect of a complex package may conflict with a personal view on approach, and the whole package or IV itself may be rejected too hastily because of this.

There can be no doubt that as the technology advances attempts will be made to use it in education. Some will see it as the answer to their problems — teacher shortages or an untrained workforce. Others will use it just because it is new. Whatever happens, the work of the IVIS project will be seen as one more step along a path of increasingly sophisticated and effective devices for teaching and training. We still need teachers, but the resources available to them and the techniques they use are changing.

3 *What's the Use of LOGO?*

Paul Ernest

The computer language LOGO is now available in one form or another for most microcomputers. Great claims have been made for the language. LOGO enthusiasts argue that it offers not only a new experience in programming but a revolutionary new way of learning mathematics. How are mathematics teachers to evaluate such claims? In this chapter I will try to offer a more sober judgment on what LOGO offers teachers and students of mathematics.

The programming language LOGO was created twenty years ago in Massachusetts by a team including its best-known proponent, Seymour Papert. In this book, *Mindstorms*, Papert makes some sweeping (and inspirational) claims for one aspect of LOGO, Turtle Geometry.[1] Although LOGO has other uses such as list processing and numerical programming I shall focus on Turtle Geometry, since the strongest claims are made for this aspect. Papert has the following to say:

> We are learning how to make computers with which children love to communicate. When this communication occurs, children learn mathematics as a living language. Moreover, mathematical communication and alphabetic communication are thereby both transformed from the alien and therefore difficult things they are for most children into natural and therefore easy ones. The idea of 'talking mathematics' to a computer can be generalized to a view of learning mathematics in 'Mathland'; that is to say, in a context which is to learning mathematics what living in France is to learning French.
>
> The kind of mathematics foisted on children in schools is not meaningful, fun, or even very useful.
>
> In order to break this vicious circle I shall lead the reader into a new area of mathematics, Turtle Geometry, that my colleagues and I have created as a better more meaningful first area of formal mathematics for children.[2]

These are bold and sweeping claims, so bold that they have the ring of an ideological and near-religious fervour. Papert is a visionary who looks ahead to a golden time when all mathematics teachers share his vision and the lack of resources and external examination syllabuses no longer constrain mathematics teaching. We need visionaries, but the sceptical teacher may well ask what LOGO can offer mathematics learners here and now. I believe that exposure to LOGO can help children's learning of mathematics in three ways, in both primary and secondary schools.

1 LOGO can help children to learn mathematics content — the concepts and skills of mathematics.
2 LOGO can help children to learn the processes of mathematics, particularly the general strategies of problem-solving.
3 LOGO encourages new learning and teaching styles, including cooperative group work, discussion and investigations.

Children Learn Mathematics Content through LOGO

In Turtle Geometry children make drawings by directing the Turtle (usually represented by a small arrowhead) around the monitor screen. This provides concrete experiences of a number of mathematical concepts and skills from geometry and other parts of mathematics.

Estimating Distance

The first problem a child encounters on learning to direct the Turtle's movements is having to decide how many units of forward movement are required to get the Turtle to its desired destination. This is usually tackled by a trial and error process of 'homing-in' on the desired endpoint. As children practise drawing with the Turtle on the screen they develop the ability to estimate screen distances in terms of the arbitrary units of length travelled by the Turtle. A number of relationships involving length are often discovered, such as the fact that in a right-angled triangle with sides A, O and H, H is longer than O and A but shorter than O + A.

Angle

The first two concepts a child encounters in controlling the Turtle are those of distance and angle. Use of the commands Left and Right require the child to experiment with and master angle measure in degrees, probably starting with 90° angles. Thus the child uses angle measure and develops angle estimation skills, as with distance. Further, an underlying dynamic conception of angle is developed, for in Turtle Geometry angle is given a concrete meaning as an amount of turn.

In addition to the development of angle concepts and skills, the child discovers the angle properties of plane figures. Thus a straight line angle is 180°, the exterior angle of an equilateral triangle is 120°, and the exterior angles of a polygon sum to 360°. This last fact Papert calls the Total Turtle Trip Theorem: in any Turtle trip which starts and finishes with the same position and heading the Turtle has turned through (a multiple of) 360°.

Shape

In drawing figures with the Turtle the child is exploring the world of shape. Although most children can recognize a square, drawing one with a sequence of moves like the following deepens a child's understanding of the concept.

```
FD 100    RT 90
FD 100    RT 90
FD 100    RT 90
FD 100    RT 90
```

In creating this sequence of instructions the child has realized that the construction of a square requires four right angles, four equal sides and that there is a four-fold repetition in the construction. Understanding this enriches the web of properties and associations of the concept of square, and lays the ground for the recognition of its fourfold symmetry. The same holds true for other shapes, such as triangles, pentagons and hexagons. Beyond the realm of polygons, the attempt to draw a circle leads to a fuller understanding of that shape too, as a shape which can be approximated by a regular n-gon, for large n.

Symmetry Transformations

The exploration of Turtle Geometry usually begins with the drawing of shapes of figures by means of sequences of instructions to the Turtle, or later by means of procedures built from these sequences. Once a shape has been designed it is natural to make copies of the shape on the screen by moving the Turtle to a new position. If the direction of the Turtle is changed, the result is a rotation. If the location of the Turtle is changed, but not its direction, then the result is a translation. Thus two of the three basic symmetry transformations arise simply and naturally in LOGO, as Figure 1 illustrates.

Figure 1. Rotated and Translated Flags

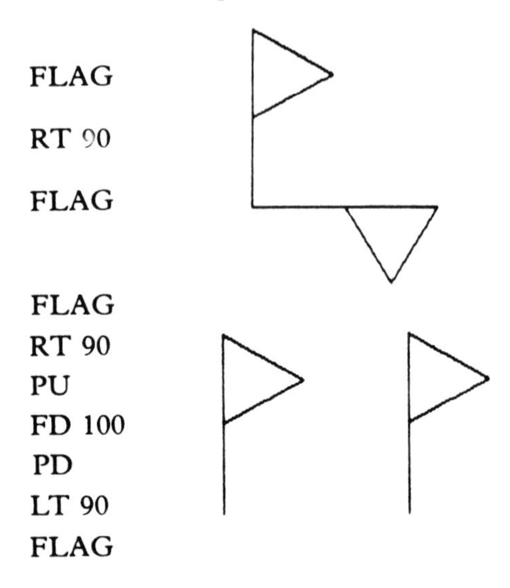

In Figure 1 a rotation and a translation are applied to a simple flag, which is generated by the following procedure:

```
TO  FLAG
    FD 50
    REPEAT 3 (FD 60 RT 120)
    BK 50
    END
```

The exploration of Turtle Geometry leads to a further development of the notions of translation and rotation. The drawing of directed lines by moving the Turtle provides a basis for the concept of a vector. Through experimentation children quickly discover the beautiful, symmetric patterns that arise from the repeated rotations of even the simplest shapes, as in Figure 2. Thus LOGO enables children to discover and apply symmetry transformations in a creative and original way.

The third symmetry transformation, reflection, is not dealt with so easily. Children do discover that the interchange of just the LEFT and RIGHT instructions in a procedure produces a mirror symmetric shape. This transformation cannot be carried out as immediately as rotation and

Figure 2. A Symmetric LOGO Design

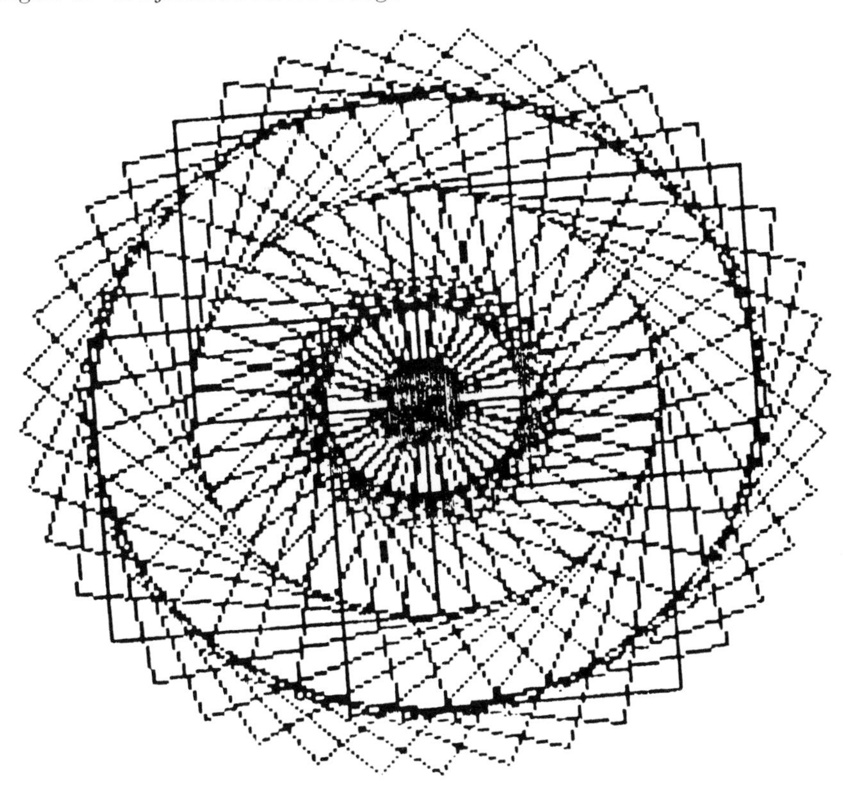

translation. It either requires the adaptation of a procedure, or can be carried out by software written for the task.[3]

Enlargement and Similarity

In making screen drawings children often wish to use several sizes of a single shape, for example, a square. After keying in instructions for the drawing of different sized squares, children are usually very receptive to the idea of a general procedure such as the following:

```
TO SQUARE      :SIZE
REPEAT 4 (FD   :SIZE RT 90)
END
```

Procedures such as this allow a child to generate a family of similar shapes. In creating these shapes the child is coming to grips with the concepts of enlargement, similarity, scale-factor and ratio, albeit in concrete form. These are central mathematical concepts which need to be experienced concretely, as in Turtle Geometry.

Variables and Algebra

Algebra is traditionally one of the most difficult areas of mathematics because of its abstraction and formality. Variable is perhaps the central concept of algebra, but unfortunately, as Kuchemann has shown, only a small proportion of children achieve a full understanding of it by the age of 15.[4] LOGO provides a meaningful context for the introduction of the variable concept. For example, the procedure TO SQUARE :SIZE listed above draws squares of any size by employing a variable :SIZE. Here the desire for a general procedure to draw squares of any size gave rise to the need to use a variable. Further, the variable is embedded in a meaningful context, and is named in a way that relates to its meaning. The power that is achieved by introducing variables provides a strong motivating factor for children.

We have seen how a variable allows a single procedure to draw squares of any size. Similarly, the use of a variable in the following short procedure draws a spiral:

```
TO SPIRAL :S
REPEAT 100 (FD :S RT 90 MAKE "S :S + 5)
END
```

LOGO is not unique in providing a context in which variables are meaningful. Any programming activity provides such a context, as Fletcher and Booth confirm.[5] However, Turtle Geometry may provide the quickest and least formal entry into programming situations in which the use of variables gives the user great power. Despite this ease of access, one major obstacle to children's easy use of LOGO variables should be mentioned: the horrible syntax involving quotes and colons, illustrated in the procedure SPIRAL.

Recursion

One of the powerful ideas implemented in LOGO is that of recursion, which allows procedures to call themselves up as subprocedures. Children use this idea as a means of initiating repetitive or iterative procedures. Thus, for example, a simple recursive procedure for drawing an enlarging nest of squares is as follows:

```
TO NEST :SIDE
SQUARE :SIDE
MAKE "SIDE :SIDE + 5
NEST :SIDE
END
```

Readers are invited to provide their own illustrations here, by trying out this procedure themselves (with some suitable choice of value for the variable :SIDE).

Recursion is not one of the topics traditionally included in the school mathematics curriculum. However, the concept of recursion does underpin a number of concepts which are common in high school mathematics. These include iteration and iterative processes; inductive and recursive definitions such as those of the factorial, power and Fibonacci functions and the Euclidean Algorithm; and proof by mathematical induction. As programming and computers increase their impact on the mathematics curriculum it is to be expected that recursion will be given more prominence in the mathematics curriculum, as is already happening with iteration.

Beyond the current mathematics curriculum the idea of recursion leads to a number of exciting areas of mathematics. The self-calling-up of procedures is very close to the idea of self-reference. This idea has played a very important role in mathematical logic, from Russell's Paradox to Gödel's Theorem and beyond. In his exciting book *Gödel, Escher and Bach*, Douglas Hofstedter uses the theme of self-reference to link the music of Bach and the art of Escher with Gödel's Theorem.

By way of endless recurrence the idea of recursion leads to the concept of infinity. Recursion in Turtle Geometry also leads to Fractals, the self-similar curves of Mandelbrot, which in turn lead to the idea of infinitesimals. Recursion in Turtle Geometry also allows the construction of space filling curves, which played such an important part in analysis in the nineteenth century.

Overall, LOGO, and in particular Turtle Geometry, can help in the learning of the content of mathematics, from the estimation of distance, angle and shape for younger children to ideas including recursion, iteration, self-reference, infinity and infinitesimals, fractals and space filling curves for older students. In addition to this range of content and skills, LOGO has a further benefit. Concepts and skills learned during LOGO programming are dynamic and hence meaningful. Consider angle. This is experienced dynamically as an amount of turn. A dynamic conception of angle is more likely to be applied correctly than the more static conception which some children acquire from protractor use. Similarly the notion of the exterior angle sum of a polygon is acquired dynamically in LOGO, as the total amount of turn of the Turtle. LOGO has the virtue of presenting many of the concepts of mathematics dynamically and concretely, in a form readily assimilated by many children.

LOGO Teaches General Problem-Solving Strategies

There are a number of ways in which programming Turtle Geometry in LOGO teaches general problem-solving strategies.

1. Children programming in LOGO learn to think algorithmically. This means that children learn to build up sequences of moves to achieve specific goals, whether their own or tasks provided by the teacher.
2. Children learn to think procedurally, to 'break large problems down into small manageable units — into "mind-size" bites'.[6] Procedural thinking allows children to determine subgoals on the way to a larger goal, and achieve these one at a time.
3. Programming on LOGO can provide experience of both 'top-down' and 'bottom-up' strategies in solving problems. A top-down strategy is like that employed in procedural thinking. A problem or goal is analyzed into a number of modules or subgoals. Each of these is further analyzed until all the subgoals are attainable. Procedures for the subgoals are then put together until the overall goal is achieved. A bottom-up strategy begins with sequences of moves or with existing procedures. These are supplemented and built upon until the goal is achieved. The top-down and bottom-up strategies are basically analytic and synthetic thought in a new guise.
4. Programming in LOGO is an activity which parallels all the stages of problem-solving in George Polya's heuristics.[7] The analogy can be shown like this:[8]

Problem-Solving	**Programming**
understand problem	analyze problem
devise plan	write program
execute plan	run program
check solution and review plan	debug and modify program

In problem-solving, after the problem is understood a plan is devised which may be analytic or synthetic. Likewise in LOGO the goal or problem is analyzed and can then be approached by a top-down or a bottom-up programming strategy. In both problem-solving and programming the proposed path to the solution, be it a plan or a program, is tried out. Finally, the success of this path is evaluated and any flaws in the solution or bugs in the program are remedied.

Evidently there is a very striking analogy between programming in LOGO and problem-solving, and both activities involve the development and use of the same general strategies. The same strategies are also evident in mathematical modelling, which involves testing a proposed solution against the situation it is intended to model. This parallels the debugging aspect of programming which is sometimes omitted in problem-solving (i.e., not checking solutions), but which is not easily left out in programming.

It has been shown that programming in LOGO teaches and fosters general problem-solving strategies. However, these are not the product of a short-term exposure to LOGO, but require a fairly sustained experience of programming. Furthermore, this benefit is not exclusive to programming in LOGO. Although entry to BASIC is not gained so easily as to LOGO via Turtle Geometry, probably the same general problem-solving strategies can be learned, ultimately.

LOGO Encourages New Learning and Teaching Styles

Since the publication of the Cockcroft Report in 1982, teachers throughout the education system have been more aware of the range of styles required for good mathematics teaching. In addition to exposition by the teacher and consolidation and practice, children need to experience investigational work, practical work, problem-solving and discussion in their learning. The inclusion of these newer elements in the classroom has been accelerated in secondary schools by the development of the GCSE examination in mathematics with its emphasis on investigations, problem-solving and practical work in the coursework projects for assessment.

LOGO programming by children is one way in which some of these needs can be met. In the previous section we saw how LOGO helps children to acquire and develop the general strategies and modes of thought which underlie problem-solving. Some further aspects are as follows.

Investigations

Investigational work depends on problem-solving, and as we have seen LOGO can offer a great deal here. Beyond problem-solving, investigations require children to explore open-ended situations and to set their own goals. Turtle Geometry provides an open, miniature world in which children set their own goals and explore for themselves. Exploration is a very appropriate term to apply to LOGO, for children can explore not only the VDU screen with the Turtle but also the hidden space behind the screen. Beyond this realm of physical exploration LOGO opens up worlds of shape, pattern, design, symmetry and more for exploration.

One of the key features of LOGO for investigational work is that children can set their own goals and pose their own problems. Thus children can feel that they 'own' their mathematics in that the goals and approaches are their own creation. In other words, by facilitating problem-posing as well as problem-solving (and freer exploration) children achieve an ownership over their mathematics. A factor which contributes to this is what has been termed 'degoaling'. This describes changing their overall goal when, halfway there, an accidental effect leads them to think of an alternative goal or problem. This type of divergent thought is an asset in investigational work and indeed in life in general. LOGO provides opportunity for this kind of creative thought which is not otherwise common in the mathematics classroom.

Discussion

Genuine pupil-to-pupil and pupil-to-teacher discussion about mathematics in the classroom is not as common as Cockcroft suggests it should be. Since it is common for children to work on LOGO projects in small cooperative groups a large amount of real discussion takes place. Celia Hoyles and her colleagues in London have recorded a great deal of rich mathematical discussion between pupils working on LOGO.

LOGO work also encourages pupil-to-teacher discussion, and often brings about a new relationship with the teacher as facilitator, collaborator and even co-explorer. Thus LOGO encourages discussion in mathematics.

Personal Qualities

As is the case with other investigational work in mathematics, LOGO programming can have a very positive effect on the personal qualities of pupils over an extended period. The achievement of mastery over the Turtle brings confidence. The perception of errors in programming as 'bugs' which are to be expected and 'debugged' also adds to confidence by diminishing the fear of failure. The setting of own goals and the ability to vary these (in 'degoaling') further adds to confidence as children learn that their own judgment is valuable in mathematics lessons. Involvement in LOGO projects with a growing sense of confidence develops persistence in children. Finally, provided that LOGO work is seen as mathematics, it should add to children's interest in mathematics and enhance their attitudes towards mathematics.

Learning

LOGO and Turtle Geometry can provide a dynamic and active learning experience in mathematics. This has been demonstrated for a number of concepts and skills in mathematics as well as for the general strategies of problem-solving and investigational work.

Seymour Papert[9] claims that the active learning brought about through LOGO is based on the theories of Piaget. However, there is an even closer link with the work of Jerome Bruner. Bruner suggests that our learning is represented mentally in three modes of progressive complexity: *enactive* representation (the memory of an active physical experience such as tying a shoe-lace); *iconic* representation (a simplified pictorial image, such as in a Russian icon); and *symbolic* representation. The learning of LOGO fits neatly into these stages. First of all, young children are best introduced to Turtle Geometry by acting out the movements of the Turtle and then by controlling the physical movements of a robotic Turtle (or the Big Trak toy). Second, children direct the pictorial Turtle around the VDU screen in direct drive. Both these stages lead to iconic representations in their minds. Third, children learn to operate purely symbolically as they write procedures in LOGO. Few areas of mathematics operate so naturally in all these three modes of representation. Even fewer manage to keep these three levels linked so naturally, thus providing meaning at the symbolic level in terms of imagery and action.

Conclusion: LOGO Is Worthwhile

I began this chapter by asking if Seymour Papert's sweeping claims for LOGO were justified. I have shown that one aspect of LOGO programming, namely Turtle Geometry, has a great deal to offer mathematics teaching and learning. It can help in teaching a wide range of mathematical content. It helps develop general problem-solving strategies. It can help to encourage new approaches to the teaching and learning of mathematics. Thus Papert's claims, although extreme, should not be dismissed in their entirety. LOGO is not, of course, a universal panacea. Learners will only get out of it as much as they put in. However, children seem to really enjoy working with LOGO. Even the very least able children with short attention spans will immerse themselves in Turtle Geometry without a break for eighty minutes. For this reason alone LOGO is a valuable asset to mathematics teaching.

Notes

1 S. PAPERT, *Mindstorms: Children, Computers and Powerful Ideas*, New York, Basic Books, 1980.
2 *Ibid.*, pp. 6, 50, 51.
3 R. GOLDSTEIN, 'Mathematics after Logo', *Mathematics Teaching*, 115 (June 1986), pp. 14–15.
4 K. HART, *Children's Understanding of Mathematics 11–16*, London, John Murray, 1981.
5 T.J. FLETCHER, *Microcomputers and Mathematics in Schools*, London, DES, 1983 (para. 104); L. Booth, *Algebra: Children's Strategies and Errors*, Windsor, NFER-Nelson, 1984.
6 F. REZANSON and S. DAWSON, 'The Logo Cult', *Mathematics Teaching*, 115 (March 1985), pp. 5–7.
7 G. POLYA, *How to Solve It*, Princeton, N.J., Princeton University Press, 1945.
8 R. NOSS, 'Doing Maths While Learning Logo', *Mathematics Teaching*, 104 (September 1983).
9 PAPERT, *op. cit.*

NEW ASSESSMENTS

A key area of innovation as we move towards the 1990s is the introduction of new forms of assessment. The *National Criteria for the General Certificate of Secondary Education* (GCSE) in mathematics were published in 1985 (SEC, 1985), and the new examination itself first introduced in 1988. The examination incorporates a number of innovative features: an emphasis on problem-solving and practical applications, and the introduction of extended pieces of work for school-based assessment. In the area of mathematics teaching, particularly that of the more academically orientated pupils, these new emphases represent a significant change of focus. Apparently this is no accident. The new examination has been explicitly represented by ministers as an assessment driven reform of the school curriculum, although mathematics has not been singled out as a special target. All of the chapters in this section make direct reference to innovatory aspects of the GCSE examination in mathematics.

A further area of innovation in mathematics is the introduction of attainment targets or 'benchmarks' for children at 7, 11, 14 and 16 years of age. This is part of the specification of the *National Curriculum* for pupils aged 5 to 16 for state controlled schools. Whilst GCSE provides the terminal assessment at age 16, tests based on the attainment targets at ages 7, 11 and 14 are being introduced from the beginning of the 1990s. The National Curriculum Mathematics Working Group (1987) provided a rationale for their approach in mathematics — welcomed by progressive mathematics teachers — and a more specific listing of attainment targets by mathematical topic areas. An overall assessment framework has been specified by TGAT, the National Curriculum Task Group on Assessment and Testing (1987), which proposes ten levels of attainment in each academic subject including mathematics. The lowest three levels are those intended to apply to 7 year olds, whilst the top four levels are intended to correspond to GCSE grades in the range A to F, for 16 years olds. One of the key strengths of the assessment framework is that the attainment targets are not strictly age related. A spread of levels is recognized as appropriate for children of a fixed age, to account for the spread of attainments, in each subject area. This is especially apposite for mathematics, in view of the seven-year spread of attainment amongst normal pupils at age 11, as remarked by Cockcroft (1982).

No discussion of innovation in mathematics assessment would be complete without a further mention of the pioneering work carried out by Derek Foxman and his colleagues at the Assessment of Performance Unit. They have developed and carried out widespread testing of innovative designs, including the assessment of attitudes, oral and practical work, problem-solving, and cooperative group work, including pupil-to-pupil discussion (not all of which is published yet), as well as large-scale achievement testing (APU, 1979–1982, 1985). One of the chapters discusses their assessment framework.

As we move into the 1990s it is clear that testing and assessment are to be key issues in the teaching of mathematics, indeed, for the whole of education.

References

Assessment of Performance Unit (1979–1982) *Mathematical Development, Primary Survey Reports 1–3, Secondary Survey Reports 1–3*, London, HMSO.

Assessment of Performance Unit (1985) *A Review of Monitoring in Mathematics 1978–1982*, London, Department of Education and Science.

Cockcroft, W.H. (1982) *Mathematics Counts*, London, HMSO.

National Curriculum Mathematics Working Group (1987) *Interim Report*, London, Department of Education and Science.

Secondary Examination Council (1985) *The National Criteria for the General Certificate of Secondary Education*, London, HMSO.

Task Group on Assessment and Testing (1987) *A Report*, London, Department of Education and Science.

4 *Classroom-based Assessment*

Susan Pirie

Uppermost in many mathematics teachers' minds at present is their worry about course work assessment. This chapter takes an overview of current practice related to assessment, and considers ways of integrating the demands of schemes such as the GCSE into classroom teaching.

One of the concomitants of anxiety is often an inability to separate important features from irrelevancies and to see clearly to the heart of the problem. An outside perspective can often help to clarify a situation which seems intractable from within. It is therefore necessary to look closely at assessment in general before focusing on teacher-based assessment. It is pertinent to consider more precisely the language to be used. 'Marking' and 'assessing' are not synonymous. They are, however, used interchangeably by many people, making it difficult to discriminate between teachers' actions and intentions. It is particularly important with the advent of attempts to assess practical and oral work that the distinction be made clear. It may sound tautologous to state that marking requires a mark to be written on a piece of work. This is usually a number, letter, tick or cross, and can be interpreted in many ways. What is crucial is that although the mark may well be the result of an assessment of the worth of the work, it may equally well be nothing more than a recognition of the work's existence, its length, or, in mathematics particularly, whether it is right or wrong. Assessment, on the other hand, implies that a considered judgment has been made on some aspect of the work. This may be communicated by a mark, but discussion or alteration to a teaching programme are also possible outcomes.

Why Assess Work?

It is worth pausing with this question since teachers frequently fall into a pattern of marking pupils' work simply because it is the thing to do. This is not a reason to be ignored. Inexperienced teachers can feel that unless they are seen to be doing a certain amount of marking their senior colleagues may judge them to be lazy or, worse, uncaring about their pupils' progress. However, unless the teacher has some other explicit reason for assessing pupils' work the effort will be little more than a time-wasting charade. Time, it is important to remember, is a commodity which teachers do not have in abundance. Basically reasons for assessing can fall into three categories: for the benefit of the pupil; for the benefit of the teacher; to satisfy the perceived needs of others. The requirements of the third category are rather different from those of the first two and will be considered first.

'Everything at school is assessed', 'the system expects it', are possibly accurate but nonetheless disturbing responses. They posit a view of schooling as 'doing, remembering and testing', rather than 'developing pupils' understanding and mental growth'. It is true that many parents see a marked piece of work as evidence that the teacher is doing his job properly: concerning himself with the work of their own particular children. The converse impression, that the lack of regularly marked work implies an indifference to the progress of the pupils, may also be held by parents in the absence of information to the contrary. Assessment procedures should be appropriately selected by teachers, not dictated by parents' needs, but parents have inalienable rights to be interested in the education of their sons and daughters, and teachers should take it upon themselves to ensure that information about their forms of assessment is available. In its absence parents have no option but to fall back on their memories of the meanings of marks when they were at school, and these may be highly irrelevant today. Other assessment demands are imposed by the desirability of grouping pupils according to their ability whether to achieve streamed or true mixed-ability classes, by the need to inform other teachers about particular pupils' progress and, more dubiously, by the schools' wish to keep records on pupils. From time to time it may be necessary to concentrate on assessment for these purposes, but the majority of the time the over-riding factors should be the value to the pupil and the teacher.

Assessing and marking children's work is one way of reporting to them how they achieved at a particular task. Many pupils expect the school system to provide them with feedback in such a form, although written comments or a personal discussion with the teacher might be a much more appropriate and effective method. Marks can be used for motivation — 'I did better than last time' or 'Rani had a higher mark then me' — although such competitiveness may have doubtful validity. Pupils are tempted to focus on the mark rather than gain understanding from reviewing their work. Marking can also offer encouragement — 'I got it right' — or pressure — 'I'd better do it 'cos I have to hand it in'. Despite pupil expectations, one of the major reasons for the day-to-day assessment of work in an educational setting must be to diagnose each child's weaknesses and strengths and then to devise some apposite action. This affects both teacher and learner, as a means of communication has to be found which apprises the learner of her difficulties and offers ways of working at the problem. If this is an aim, marking is rarely of value.

It is possible, even desirable, that from time to time a teacher's reasons for assessing work take on a more egoistical hue. To find out what the class can do, to find out what it cannot do, to evaluate a piece of teaching are all legitimate reasons for assessment for which marking has no relevance whatsoever and considered teacher action is the appropriate response. In order to use time wisely and economically it is essential to address the question of 'why assess?' with seriousness and honesty. The later part of this chapter looks at realistic ways to react when 'why' has been established.

What to Assess?

Even for the traditional piece of written mathematical homework this should not be a trivial question. Accuracy, results, method of working, evidence of thinking, mathematical ideas assimilated, transfer of understanding, demonstration of skills and techniques, personal effort, all these are responses which should trigger different approaches and outcomes of assessment. Coursework of an extended, practical or investigational nature poses additional criteria to be considered. Much of this type of work involves pupils in activity, discussion and rough working. Following this they are usually asked to produce a written account of some form. The value of this last activity is discussed later.

Should observed activity be assessed? What about overheard or teacher-led discussion? Even when considering the written work there are different areas to consider. For exploratory and problem-solving approaches to the learning of mathematics to be effective for pupils it is critical that the image of mathematics as 'a neat piece of written algebra which is either right or wrong' be eroded in the minds of the pupils and of their teachers. The real search for mathematical truths involves conjectures and predictions, checking and refining, errors and reappraisal and much rough working. To encourage this view of mathematics a distinction needs to be made, overtly, between recording and writing-up — both valuable activities but with different purposes. Recording is personal. It is to record passing thoughts, hunches and ideas. It is to try out calculations, illuminating diagrams and possible lines of attack. It is for making mistakes and building on the insight thus obtained. It is not about neat writing, erasing unfruitful attempts, ruled margins, right answers and external coherence. Writing-up, however, is for communication to others. It may not be necessary to explain the false starts, unprofitable avenues and inaccuracies that occurred during the course of the work. The pure mathematician may present a few lines of elegant proof as the result of years of search and struggle. It may, on the other hand, be of value to expose these events to the reader so that others may be made aware of fruitful and frustrating paths of exploration, and may themselves build on the author's trains of thought. This is the writer's decision. What is necessary is a clear explanation of the problem and comprehensible account of the activity and its outcome, and this communication skill probably needs to be explicitly taught. What then should be assessed? Clearly one may not assess the recording, although the recordings might be consulted to throw light on what mathematical thinking was taking place. It may be born in mind that only positive assessment can be made thus; absence of written evidence of thinking is not evidence of lack of thought. The final (or concurrent) writing-up can be judged from two stances. Is it a clear, informative account of the mathematical activity? Alternatively, what is the level of mathematical activity as revealed by the write-up? It is this latter aspect which is most commonly the basis of assessment, and if this is to continue to be so, pupils should be made aware of the processes which need to be explicitly articulated in their writing. The hardest aspect of all to assess is that which should be the most valued: has an increase in mathematical understanding taken place?

The aim in the preceding paragraphs has been to raise to a level of consciousness the decisions which ought to be considered before selecting an appropriate method of assessment; only then can an answer be given to the question, 'How to assess work?'

Formative or Summative Outcome?

Assessments can be divided into two categories: formative assessment and summative assessment. The former is a judgment made with the express intention that action will result, usually in the form of communication between teacher and pupil, and the subsequent offer of relevant follow-up tasks. The latter implies a judgment passed on mathematical ability at a particular moment and assumed to be a comment on the current position of understanding reached. Two illustrations of these categories are, respectively, considering a pupil's homework with a view to planning the next lesson and marking end-of-year examinations. Newly added to this last category is the assessment by teachers of coursework for GCSE. If this is not to become the time-consuming burden which many teachers fear, it is necessary to look closely at existing assessment and marking and prune away self-imposed, unnecessary or inefficient tasks.

Formative and summative intentions require very different assessment procedures. Consider first ordinary classwork and homework; answer the questions 'why?' and 'what?' If the responses concern mark-book records, parents' evenings, a picture of the class as a whole, where pupils are at the end of a topic, then a quick, briefly recordable, summative scheme is needed. Common currency of meaning among colleagues may also be desirable. Two suggestions are offered here for serious consideration. The first is to rank order the pieces of work as they are read without heart-searching over precise, individual positions, and accepting that only accuracy within, say, five places, is expected. Then number the pile from 1 to n. This achieves a quick overview of pupils' relative understanding or ability on a particular piece of work. Initially this method requires discipline and strength of purpose and conviction that precise rank ordering is not necessary, but the rewards in time saved can be considerable. It may well be undesirable to give these marks to the pupils who will probably try to attach too much importance to their exact positions, but bear in mind that feedback to pupils was *not* the reason for the assessment on this occasion. An alternative and possibly easier method is, having read a piece of work, to allocate it to one of four groups — A, B, C and D. Do not agonize over A−− and B++, because the purpose is to produce a quick, rough, descriptive mark. The groups are unlikely to be of equal size, but will give a view of the spread within the class on that specific topic. If the purpose of the homework was practice of a particular technique, why mark the work outside the classroom? A tick acknowledges the receipt of the work, and pupils marking the work in class takes but a few minutes.

Formative assessment clearly needs more attention to the detail of individual pieces of work. Time spent on this task aims to facilitate the learning of mathematics. A single mark has little meaning and comments, whether written or discussed with a pupil, are more appropriate. If the purpose is to inform future teaching, then perhaps it is unnecessary to communicate with individual pupils. The feedback can be in the form of a general class statement.

All the foregoing suggestions have been made relative to traditional, written work. How does practical, investigative and group work, given public credibility by the Cockcroft Report (1982), affect the scene? Some teachers have been working in this way with their pupils for many years. It is, however, a sad, but nonetheless true, indictment of British educational society that 'if it is not seen to be assessed, it is not seen to be valued', and many teachers have, with arguably the best interests of their pupils at heart, refrained from 'wasting time' which could be more profitably spent concentrating on the examination syllabus. Enjoyment will never be an item assessed for external purposes! Course work, however, now is, and suddenly teachers are being asked to work in unfamiliar ways and at the same time summatively to assess their pupils' performance in these new areas of learning. What is crucial at this moment is that teachers retain their confidence in their own professional capabilities. They know their pupils' real mathematical abilities far better than any external examiner can ever do when basing a judgment, which may dramatically affect a candidate's future, on a couple of written, timed papers. Teachers must stake their professional integrity on their ability summatively to assess their own pupils fairly and accurately. If pupils of any age are to approach mathematical learning successfully by new routes, they will, in addition, need guidance through formative assessment. They will need continual feedback, encouragement and assistance.

What Does Assessing Coursework Involve?

Again the question 'what is to be assessed?' must first be asked, the answers discriminated one from another and the pupils made aware of what is expected of them. In a practical task, is the final artifact or the 'doing' to be judged? Are planning and research important or are only physical skills to be assessed on this particular occasion? Is the aim of the practical work to enhance mathematical learning in general, or to develop specific skills and techniques? The aim behind any project must affect the light in which it will be judged. How far should choice of suitable materials for a given task be left to the individual pupil, particularly if this radically affects the possible outcome of a task? This last question opens up the whole minefield of help which is at the root of teachers' fears as to their ability to make universally valid assessments; how much guidance may pupils be given? Careful consideration of what is being assessed will help to resolve this dilemma. Consider, for example, the task to build a scale model of the classroom. If the final model is the focus of assessment, then discussion of the best scale to use is quite in order, since the judgment will be passed on the outcome of following this advice. If, however, the underlying mathematics is at issue, then tips on the best glue and cardboard to use for the construction are totally permissible since the beauty of the final model will enhance a pupil's feelings of achievement without prejudicing the assessment of the work. A third possible reason for this practical task might be to test measuring skills, in which case neither help on choice of scale nor model building hints will affect the area under consideration. In such an extended piece of work the teacher may wish to look at several different aspects of activity and a form of continual assessment is apposite; a combination of summative and formative methods should be used. In the above illustrative example the ability to measure accurately and use appropriate tools could be judged, and then the measurement discussed with the pupils so that inaccuracies here do not affect the final model. The use of scale could then be considered for suitability and accuracy of calculation, and any errors fed back to the pupil in order that subsequent assessment of model building skills should not be jeopardized. This kind of approach is in line with the common examination practice of following through and awarding marks for written work based on an earlier incorrect calculation.

Other forms of newly assessible coursework include investigations and problem-solving. One of the aims behind the inclusion of this type of open working in the experience of school mathematics is to increase pupils' enjoyment and remove the artificial view many have that to answer any problem in a mathematical lesson they need consult only the necessary and sufficient information given in the question; irrelevant detail will not be offered and assumptions or predictive approximations will not be required. Formal assessment of open working must be approached with caution, particularly when it is not topic content but use of mathematical processes which is being examined. One of the advantages of investigative work is that a pupil may explore paths of personal interest, thus making mathematical activity relevant to the individual's current abilities and inclinations. The dangers for assessment are that pupils may not create for themselves the opportunity to display their full capabilities. Should pupils be expected to choose an activity which will demonstrate as many skills as possible, as in some home economics examinations, or is this the responsibility of the teacher? Is one of the areas of assessment to be the ability to predict a mathematically rich avenue of enquiry? It is in this field that assessment by a teacher who knows the pupil is essential. Over a span of time and breadth of experiences, professional judgment can be made which does not depend on a pupil attempting to reveal, in the exploration of one problem, all her capabilities.

How Can Coursework Be Assessed?

This question can now be addressed from the two angles of day-to-day classroom working and formal, final accreditation. Unless the sole focus of attention is to be the end product, be it model, write-up or correct answer, much of the assessment must take place in the classroom with the teacher eavesdropping on the pupils talking, engaging them in discussion about what they are doing, answering a question with a question and really comprehending the reply. Pupils need encouraging to talk about mathematics before being expected to write about it. There need be no attempt to evaluate each pupil's contribution in each lesson. A more realistic strategy is to approach the class with an intention to focus on four or five individual pupils and assess their working on some pre-arranged aspect. It can be valuable to inform the pupils about what you will be focusing on as this is seen as an important aspect of their mathematics. At the end of the lesson a few sentences jotted down help to produce any more formal record which might later be needed. A more structured and probably easier method of classroom assessment is to use a checklist of areas to consider and a record card for each pupil which is filled in from time to time. One example of this, which has been used with success, is described in detail in Pirie (1987). It offers a way of recording which enables the teacher both to be systematic and to respond to unforeseen insights when they occur.

A written account of a piece of mathematical activity is a new concept in most classrooms, yet if pupils are to form some record of their progress in open working situations it is a valuable adjunct to their learning. It will not, however, come easily to most children because it does not lie within their expectations of the mathematics lesson. Written mathematics for most pupils involves numbers and letters, but not words. Many pupils when asked to write down the explanation they had just given orally would join with Simon, an articulate, bright, motivated 11-year-old, in saying, 'But I don't know how to write it in maths.' It is wise, therefore, sometimes to have, as the aim of a particular mathematical task, the assessment of the write-up, not for the thought and understanding revealed, but for its effectiveness as a means of revealing any learning which might have taken place. This is not to say that every project must be written-up. Class discussion, peer interaction or practical demonstration may frequently be better ways of concluding an activity. Ideas and expressions of understanding can be lost in the struggle to provide a written account. Talking allows modification of thoughts as they are articulated and is a dynamic, personally involving process within which errors can be perceived and acted upon. Writing can distance the thinker from the thought, producing a static, final record. However, nationwide credibility of qualifications is of importance in our society, and the write-up is highly valued for its very permanence and assumed irrefutable evidence of ability. Since this is the present situation, pupils should be specifically and overtly taught the skills of representation and communication which will enable their work to be justly assessed.

What about GCSE?

One thing is very clear: if pupils are not to spend a large part of the fourth and fifth years producing work from which they learn nothing, the tasks for the GCSE must be both summatively and then formatively assessed. In the late 1970s, when N and F levels were being considered as an alternative scheme to A-level, the Association of Teachers of Mathematics produced a booklet (ATM, 1978) describing various methods of assessing sixth form investigative and extended work. Much of their thinking could be used to form a basis for formal assessment of coursework at 16+. Pirie (1987) offers three alternative ways to approaching this task, all of which have the merit of relative simplicity and do not try to force investigational work into a mould dictated by an examination mark scheme. The new examination boards have each produced schemes indicating markedly different attitudes to coursework assessment, ranging from the Northern Examining Association, whose material contains statements such as 'Centres are responsible for deciding the nature of the coursework', 'teachers are free to devise their own assignments', to the Midland Examining Group, which specifies the number, length and topic area of the assignments, which must then be assessed according to a very tight numerical scheme. There are teachers who will welcome the freedom to assess coursework as they consider appropriate; there are teachers who will feel more confident working within a strictly prescribed mark scheme. The latter must be wary lest the coursework be too tightly controlled, reducing what should be creative work almost to the status and form of a series of written, timed examinations. If teachers are to assess their own pupils' work, then they should be offered guidance, but nothing more. How can work produced to fit a rigid scheme be called 'open'? Such schemes, too, can become overbearingly time-consuming if marks of 1 and 2 must be accumulated to produce a final grade of, say, B. Teachers must be trusted to make qualitative statements based on their knowledge of a pupil and her work. After all, university tutors have for years been able to say with unchallenged confidence, 'that is a first class piece of work; that is barely worth a IIii'. Now, as possibly never before, the classroom teacher is in a position to influence national assessment methods and demonstrate a professional competence.

The final word in the debate on how coursework should be assessed lies at present with the teacher, and it is an opportunity which should not be missed. Public assessment can be used as a means of enhancing the schoolchild's classroom experience in the field of mathematics; if teachers so choose, it need no longer be an end in itself.

References

Association of Teachers of Mathematics (1978) *Mathematics for Sixth Formers*, Derby, Association of Teachers of Mathematics.

Cockcroft, W.H. (1982) *Mathematics Counts*, London, HMSO.

Pirie, S.E.B. (1987) *Mathematical Investigation in Your Classroom*, Basingstoke, Macmillan Educational.

5 *Developments in Assessing Mathematics*

Paul Ernest

General Background

In mathematics we assess the attainment of mathematical objectives. In the past these objectives have tended to be at least partly implicit. Currently there is a move towards more explicit statements of objectives.

If we focus on the cognitive objectives of mathematics teaching, that is objectives concerned with knowledge and understanding, then a two-way classification of objectives is appropriate. There are mathematical content objectives, which are concerned with mathematics subject matter, and there are also mathematical process objectives, which concern mathematical skills, processes, abilities and strategies.

We consider first mathematical content objectives. Traditional assessments of mathematics employed a content-based classification of objectives. This is evident in the three test papers testing proficiency in arithmetic, geometry and algebra, respectively, in the traditional GCE examination in mathematics at O-level. More recent and current assessment systems have also used content-based classifications of test objectives. The CSMS project has assessed mathematics understanding in ten topic categories.[1] The Assessment of Performance Unit has developed a testing framework which employs an analysis of mathematics content into the following five areas: number, measures, algebra, geometry and probability and statistics.[2] A comparable large-scale testing of mathematics achievement by the NAEP in the US uses a similar five-fold classification of content.[3]

A contrast is provided by the SLAPONS scheme which provides a much more detailed analysis of content.[4] This scheme analyzes purely numerical skills into eighteen content objectives. Twelve of these objectives consist of the four arithmetical operations applied to whole numbers, (vulgar) fractions and decimal fractions. The attainment of a candidate on each of the objectives is reported separately, to give an overall profile of numerical skills. Figure 1 illustrates a possible individual profile.

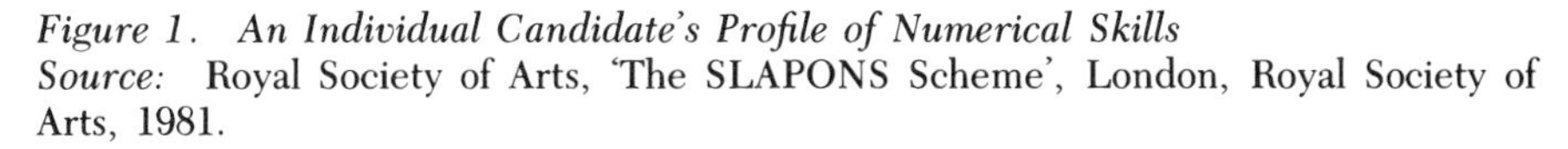

Figure 1. An Individual Candidate's Profile of Numerical Skills
Source: Royal Society of Arts, 'The SLAPONS Scheme', London, Royal Society of Arts, 1981.

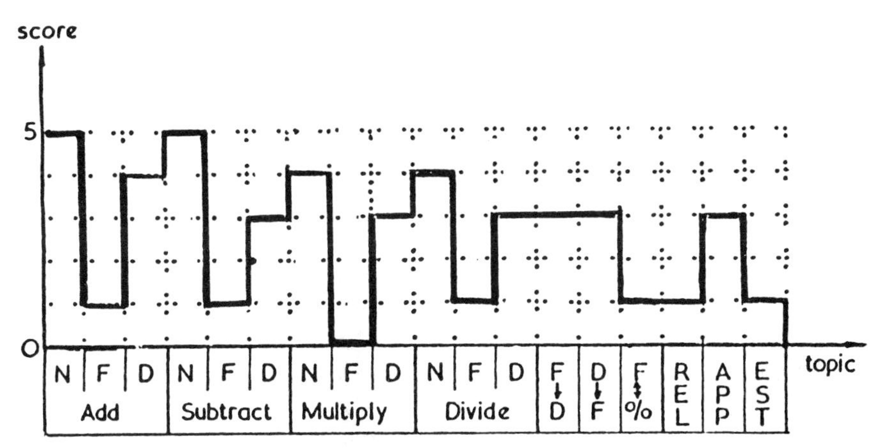

The second type of objective assessed in mathematics is the process objective. Process objectives concern the acquisition, development and employment of mathematical skills, processes, abilities and strategies by students. Process objectives have been contrasted with content objectives. The temptation to contrast process with product should be resisted, for often attainment of process objectives is evaluated in terms of their products or outcomes. Thus investigations and projects, for example, are usually assessed by means of a product, a written report.

Process objectives were first classified in the taxonomy of Bloom into the six components: recall, comprehension, application, analysis, synthesis and evaluation.[5] More recent classifications of process objectives have been designed specifically for mathematics. The NAEP classifies mathematical processes into four areas: knowledge, skill, understanding and application.[6] The Caribbean CXC mathematics examination system uses a three-fold classification of mathematical process: computation (technically: recall), comprehension (technically: algorithmic thinking) and reasoning (technically: open search or problem-solving). In addition to an overall grade the CXC mathematics examination provides a profile of candidates' performance in each of these three process dimensions.[7] A fourfold classification of mathematical processes into computational operations, pattern recognition, logical reasoning and symbolic manipulation is used in the MACOT profile.[8] The secondary mathematics assessment framework of the Assessment of Performance Unit employs two mathematical process clusters: concepts and skills, and investigating and problem-solving strategies.[9] In contrast the Mathematical Association distinguishes as many as seventeen mathematical process objectives in the cognitive do-

Figure 2. National Assessment of Educational Progress: Framework for the Mathematics Assessment
Source: T.P. Carpenter *et al.*, *Results from the Second Mathematics Assessment of the National Assessment of Educational Progress*, Reston, Va., National Council of Teachers of Mathematics, 1981.

	CONTENT				
PROCESS	A Numbers and Numeration	B Variables and Relationships	C Shape, Size and Position	D Measurement	E Other Topics
I Mathematical knowledge					
II Mathematical skill					
III. Mathematical understanding					
IV. Mathematical application					

main alone.[10] These examples serve to illustrate the variety and number of different process objectives for mathematics considered by various authorities and testing agencies.

Mathematical process and content objectives can be combined into a two-dimensional array to give an assessment framework or scheme. The framework for the NAEP mathematics assessment is shown in Figure 2 and that for the APU secondary assessment in Figure 3. Assessment frameworks of these types can be a useful aid in designing assessments as they indicate the range of combinations of content and process objectives for which test items could be designed.

The APU framework is remarkable in that it has two additional dimensions. The first consists of the context or domain of the mathematics, be it purely mathematical, mathematics in an everyday context or mathematics in other subject contexts. The second new dimension concerns the mode of assessment, which can be in a written test form or in an oral and practical test form.

A further feature of note in the APU framework is the analysis of the two process classification clusters. The analysis of the concepts and skills cluster is perhaps controversial. Three process subcategories are applied to the number and concepts and skills combination, namely: numerical concepts, skills and applications. The remaining content areas are subdivided into content, and not process subcategories. This approach could be criticized as incoherent. It suggests that the APU does not accept that there are component processes of the concepts and skills cluster which can be applied across the full range of content objectives.

The investigating and problem-solving strategies cluster is further analyzed into more specific process objectives. There is the investiga-

Figure 3. Assessment Framework for Secondary Mathematics
Source: Assessment of Performance Unit, *Mathematical Development*, Secondary Survey Report, No. 3, London, HMSO, 1982.

MODE OF ASSESSMENT: Outcomes can be assessed in written or practical test forms.

Outcome	
Attitudes	Attitudes to practical mathematics: willingness to handle apparatus; confidence; anxiety about success; verbal fluency; persistence
	Enjoyment, utility and difficulty of specific mathematical topics in the curriculum
	Liking, difficulty and utility of mathematics as a subject
Problem solving strategies	Investigations/Creativity: to be developed
	Processing information Formulating problems Strategies and methods of solution Generalising solutions Proving Evaluating results

Outcome: **Concepts and skills**	Content
Concepts	**Number**
Skills	**Number**
Applications of number	**Number**
Unit	**Measures**
Rate and ratio	**Measures**
Mensuration	**Measures**
General algebra	**Algebra**
Traditional algebra	**Algebra**
Modern algebra	**Algebra**
Graphical algebra	**Algebra**
Descriptive geometry	**Geometry**
Modern geometry	**Geometry**
Trigonometry	**Geometry**
Probability	**Probability and Statistics**
Statistics	**Probability and Statistics**

CONTEXT: The content can be set in Mathematical, Everyday or Other subject contexts

tions and creativity category, which is not yet developed. There is also the six-fold specification of problem-solving strategies as: processing information, formulating problems, strategies and methods of solution, generalizing solutions, proving and evaluating results. In making this analysis the APU has made an initial analysis of some of the specific processes involved in mathematical activity. In the next section this theme will be taken further.

The Present Context

In January 1985 the Secretaries of State of Education and Science in England and Wales published National Criteria for Mathematics.[11] These provide a mandatory basis on which the new GCSE system of examinations in mathematics is to be built for students of 16+ years of age.

The National Criteria for Mathematics have two dimensions: the mathematical content and mathematical process dimensions. The content dimension is specified in the form of two lists of topics. These lists are to provide the minimal contents for syllabuses for assessment at three levels, and are evidently based on the foundation list of topics in the report of the Cockcroft Committee.[12]

The National Criteria also contain seventeen mathematical process objectives on which candidates' attainment will be assessed. The list of processes constitutes a carefully thought out and far-rearching specification of mathematical skills. Two central themes may be discerned in the list. First, there is an emphasis on the skills involved in applying mathematics in a variety of situations including, but not limited to, everyday situations. Secondly, there is also an emphasis on the skills involved in solving problems mathematically. These two themes can be seen to underpin several of the process objectives.

An innovation of particular interest is the inclusion of two process objectives in the list of seventeen which require means other than time-limited written examinations for their assessment. These two objectives are as follows:

> . . . the ability of candidates to: . . .
>
> 3.16 respond orally to questions about mathematics, discuss mathematical ideas and carry out mental calculations;
>
> 3.17 carry out practical and investigational work, and undertake extended pieces of work.

It is evident that the specification of these seventeen process objectives, and indeed the formulation of National Criteria for Mathematics as a whole, is intended to affect the teaching of mathematics. The influence of the celebrated paragraph 243 of the Cockcroft Report is obvious.

National Criteria stipulate that discussion, practical work, problem-solving and investigations are to be assessed. For this assessment to be effective, it follows that discussion, practical work, problem-solving and investigations will need to complement the presently more common modes of expository teaching and individual practice work by pupils.

In designing an assessment scheme in accordance with the National Criteria it might prove useful to construct a framework of the type illustrated in Figures 2 and 3. The result will be rather more elaborate than those illustrated, since there are seventeen process objectives, and the content lists have not been neatly packaged into labelled topics. Nevertheless, such an exercise would help prevent the under-assessment of some of the more novel process objectives.

One of the consequences of the new GCSE examinations in mathematics is that teachers will be involved in school-based assessments of two types: first, the assessment of pupils' oral work and practical work carried out during the assessment; secondly, the assessment of pupils' written accounts of extended work on projects, problems or investigations. These two modes of assessment will be novel to many teachers. The literature on these novel modes of assessment is relatively limited, there is a need for discussion and development in this area. Given this need, I offer no apology for the tentative nature of the following proposals.

Oral and Practical Assessment

Oral and practical assessment in mathematics consists of the assessment of a pupil by means of his or her spoken response or physical performance of a task. It is evident that there are going to be difficulties and expenses in introducing assessment of this type into mathematics. So why is this a mode of assessment to be valued? What advantages does this mode of assessment have over conventional modes? I would like to suggest two sources of advantage: the fact that the candidate communicates without the use of *written* language or symbolism, and the flexibility of the situation.

The advantages of avoiding written language are three-fold.

1. A candidate may respond directly or enactively to a physical task, such as measuring a curved line. Thus we can ascertain *directly* that the candidate has certain skills.
2. A candidate can respond verbally to a question, allowing us to test objectives without the mediation of written language and the pupil's writing skills.

   These two advantages are particularly important for less able candidates. However, an advantage for all candidates is the following.

3 A candidate will find it easier to operate at a high cognitive level; to make judgments, justify, evaluate and express opinions. Candidates are used to operating at this level in verbal interactions, but not in written form, *especially in mathematics*.

There are two advantages due to the flexibility of the oral testing situation.

1 The examiner may prompt, stimulate or guide the candidate if needed. This flexible response means that we can credit the unaidedly successful candidate without failing the candidate who can be successful with prompting.
2 The examiner can focus on the individual strengths, skills and interests of the candidate. Thus, for example, in discussing a project a candidate can communicate some of his or her enthusiasm and interests.

Given these considerable advantages, the question of how candidates can be assessed in the oral modes arises. Two distinctions can be applied. First, candidates can be assessed singly or in small groups. Secondly, the focus of the assessment situation can be a task set then and there by the examiner, or the discussion of a completed piece of work, typically a project, previously completed by the candidate.

Combining these two distinctions results in a proposed four-fold scheme for oral assessment, illustrated in Table 1. These four suggested forms of assessment, which are not intended to be exhaustive, are considered more fully below.

1 Candidates can be assessed on a one-to-one basis in a structured interview situation where they respond to the examiner's tasks or questions. Extensive developmental work on this mode of assessment has been carried out by Sumner, Kyles and Sumner and the Assessment of Performance Unit.[13] The pattern followed in developing these tests is as follows:
   A The selection of test objectives.
   B The construction of oral question sequences on the objectives. The preparation of detailed response sheets to record pupils' responses and the aid given.
   C The administration of the tests and the marking of the responses to show unaided success, success with or without aid, lack of success.

   An example of a test sequence on the topic of length, from Assessment of Performance Unit,[14] is shown in Figures 4a–4c. Figure 4a lists the apparatus used which includes a sheet *A(B)* with the diagram of a circle with an inscribed horizontal diameter. The responses of a particular candidate are shown as noted by

Table 1. Forms of Oral Assessment in Mathematics

Candidate grouping	*Focus of assessment*	*Form of assessment*
	Examiner-set task	Previously completed project
Individual candidate	Individual performance on task (aided or unaided)	Discussion of project Evaluation Suggested extension
Small group of candidates	Individual contributions to group work on task (aided or unaided)	Discussion of individual contributions Overall evaluation Individual extension suggestions

the examiner. Also shown are codes for the pupil's response (*S:* successful, *U:* unsuccessful) and tester's response (*P:* prompt, *T:* teach, *N:* not given).

The advantages of this approach are that it permits a sustained assessment of a limited number of objectives, and that it should be easy to train examiners to administer and mark oral assessments of this type.

2 Candidates can be assessed individually on a completed piece of their work. In this mode of assessment a candidate can be questioned on a completed project, on a model he has made, or on the report of a mathematical investigation. The candidate can be directed to discuss aspects of his work, for example, to describe difficulties in completing it, to evaluate sections of it, and to suggest extensions and improvements of it. A checklist or profile of desired mathematical processes, skills or abilities can be used to evaluate the candidate's responses.

The advantage of this form of oral assessment is that it complements and 'rounds out' the assessment of a candidate's extended mathematical work. It also permits a candidate to be assessed in an area of his or her choice and strength.

3 Candidates can be assessed orally in small groups, on a previously completed group project or on an examiner-set joint task or on individually set tasks. This is the most tentative of the modes of oral assessment suggested here. Much of the previous discussion of oral modes of assessment continues to apply.

The difficulty of employing this mode of assessment is in distinguishing the individual contributions of the candidates and in apportioning credit accordingly. This difficulty makes it most appropriate for the usual mathematics teacher to apply this mode of assessment to his or her pupils. This also enhances the gain in candidate confidence and ease which derives from assessing candidates in peer groups.

This concludes the tentative proposals for introducing oral or practical modes of assessment into the mathematics curriculum. Further variations may be developed by looking beyond the usual scope of mathematics assessment to diverse fields such as psychology, for clinical interview techniques, or to management studies, for occupational interview techniques.

The Assessment of Extended Pieces of Work

An immediate problem arises in assessing extended pieces of student work. There are no right or wrong answers.

Extended pieces of work in mathematics can take a variety of forms. These include projects requiring the gathering of data, 'real' problem-solving, mathematical modelling, the construction of physical models, projects on aspects of mathematics such as its history, and the investigation of mathematical situations or starting points. These and other forms of extended pieces of mathematical work will be referred to simply as projects and investigations.

Projects and investigations are important because they foster sustained and independent study by pupils, and because they encourage creativity, problem-solving and other mathematical processes under-emphasized elsewhere in the curriculum. These virtues are stressed elsewhere, for example, by the Cockcroft Committee and Ernest and indeed in the National Criteria: Department of Education and Science.[15] Since the importance of project and investigations lies in the mathematical processes they foster, it seems appropriate to assess the written reports of projects and investigations in terms of the attainment of process objectives.

Two criteria which are applied in judging the effectiveness of a test are its validity and its reliability. Since the aims of encouraging projects and investigations are to nurture and develop pupils' skill in applying mathematical processes, a valid mode of assessment should test process objectives. An assessment scheme specified in terms of mathematical processes which can be applied uniformly across the range of content objectives also benefits in terms of reliability, for the lack of reliability which can arise from attempts to coordinate and compare different content-based assessment schemes may be avoided.

This discussion provides a strong justification for assessing projects and investigations by the attainment of process objectives. It should also be recalled that this mode of assessment is intended to complement more conventional time-limited written examinations, which tend to emphasize content objectives, and indeed are the most suitable way of ensuring a broad content objective coverage.

Figure 4a. Oral Question Sequence on Length
Source: Assessment of Performance Unit, *Mathematical Development*, Secondary Survey Report, No. 1, London, HMSO, 1980.

EXAMPLES OF COMPLETED TOPIC SHEETS

MATHEMATICS MONITORING - SECONDARY PRACTICAL pupil number ____________

TOPIC A

Length Measures - straight and curved lines
Geometry - rectangle and circle

Apparatus: sheet with rectangle A(A)
sheet with circle A(B)
ruler
pencil
scrap paper
string)
scissors) keep concealed until requested.
tape measure)

1. *"What is the perimeter of a rectangle?"*

	S	Sum of the sides

2. Present sheet with rectangle A(A)
"Could you show me the perimeter of this rectangle?"

	S	✓
If necessary, teach.		

3. *"How would you measure the perimeter of a rectangle?"*

		Measure each side and
	S	add them all up

If necessary, prompt for full procedure.

	N	
If necessary, teach.		

SUPTN 10.15

Figure 4b. Oral Question Sequence on Length, Continued
Source: Ibid.

EXAMPLES OF COMPLETED TOPIC SHEETS

4. *"Estimate the length of the perimeter of this rectangle."*
(7 x 4·3 cm, perimeter = 22·6 cm.)

	S	≈ 22 cm

5. Present ruler
"Check your estimate by measuring."

		22·5 cm. 7 x 2 = 14
How		Oh. Can I use mm ?
Yes		4·3 x 2 = 8·6
	S	So 22·6

6. *"What is the circumference of a circle?"*

	S	All the way round the outside

7. *"What is the diameter of a circle?"*

	S	Line through centre straight across the circle

8. Present sheet with circle A(B).
"Show me the circumference of this circle."

	S	✓

9. *"Show me the diameter drawn on this circle."*

	S	✓

SUPTN

10.16

Figure 4c. Oral Question Sequence on Length, Continued
Source: Ibid.

EXAMPLES OF COMPLETED TOPIC SHEETS

10. *"Estimate the length of the circumference of this circle."*
(~22 cm)

	U	≈ 15 cm

11. *"What could you use to check your estimate?"*
(String/π)

	S	String

If no response, prompt for string.

	N	

12. *"Do it."*

	S	Measures half
		then x 2
		(10 x 2) = 20 cm

13. *"Is there any other method?"*

	S	Use 2πr

If pupil does not suggest using (C = πD), prompt
"Would it help to measure the diameter of the circle?"

	N	

SUPTN

10.17

Various schemes of assessing projects and investigations have been proposed.[16] These schemes typically consist of a list of five to seven process objectives, each with its apportionment of marks. In some of these schemes the award of marks for each objective is criterion-linked.

The International Baccalaureate assessment scheme awards marks in five process-outcome categories:[17]

1 data collection (subject of analysis);
2 skills acquisition (means of analysis);
3 analysis (mathematical development);
4 presentation (writing up the whole report);
5 originality and difficulty of the project.

Each of these five categories is equally weighted and assigned a maximum of four marks (criterion-related). The scheme permits allowance to be made for assistance by the teacher (a maximum of two marks may be deducted from categories 1, 3 and 4). The strength of this scheme is its generality. It can be applied across the whole range of mathematical projects and investigations. A concomitant weakness is that it fails to refer to specifically mathematical processes.

In the Association of Teachers of Mathematics a seven category marking scheme for individual investigations is specified.[18] This is used to provide a profile of attainment in the following process objectives:

1 capacity to conduct and evaluate own work;
2 communication skills;
3 knowledge of mathematical concepts and skills;
4 ability to generalize, extend and formulate problems;
5 accuracy and rigour of argument.

This is more specifically mathematical but may be criticized for not distinguishing mathematical objectives sufficiently.

A scheme I use for assessing investigations is the following. The maximum number of marks that can be awarded to each of the processes is indicated in parentheses.

1 Generation and organization of data (6).
2 Conjecturing relationships (4).
3 Attempting to justify (verify, falsify, prove) (4).
4 Ability to generalize and extend investigation (3).
5 Ability to leave open questions, new starting points for investigations (3).
6 Ability to present, write up (3).

This scheme is designed specifically for mathematical investigations, as opposed to other sorts of mathematical projects. Thus it reflects a view of the nature of mathematical investigations. It may be criticized for distinguishing the categories 4 and 5. A similar scheme without this distinction, but which separates the generation and organization of data (here combined in 1) is presented in Ernest.[19]

In the construction of an assessment scheme of the types considered two variables can be manipulated. First, there is the choice of process objectives. These can be drawn from sources such as the above schemes or from the previously mentioned lists of process objectives. A further source is the six-fold analysis of mathematical processes in Bell *et al.*: generalization, explanation, proof, representation, including symbolization and the use of diagrams, classification, interpretation.[20] In addition to the various process objectives considered, an oral component can be included.

A problem arises concerning the choice of process objectives in constructing an assessment scheme. As the mathematical processes become more precisely defined, so the resulting assessment scheme becomes less widely applicable. Thus my scheme given above includes processes (4 and 5, for example) which make it unsuitable for the assessment of mathematical model building or historical projects. One strategy for overcoming this problem is to specify a number of processes (eight, for example) in the scheme, and assess student work by means of the most appropriate six process objectives. An assessment scheme of this type is proposed in Hedger and Kent.[21] In this scheme the five most appropriate process skills from the following list are employed: independence, practical manipulation, evidence of investigational approach, discussion of mathematical ideas, ability to carry out mental manipulation, and a possible as yet unspecified seventh skill. Each of the skills is given an equal weighting, so that a free choice from the list is possible. This leads us to a further feature of choice.

The second variable which can be manipulated in constructing assessment schemes for projects and investigations is the weighting given to different process objectives. This reflects the importance attached to the different objectives as well as students' ability to master them.

The various proposals for schemes of assessment for extended pieces of mathematical work share a common subjective element. By focusing on the attainment of specific process objectives, and by criterion referencing the assignment of marks for each of these objectives, the subjective element can be reduced, but not eliminated. This subjective element notwithstanding, a reliable rank ordering of students' work can be achieved.

Whatever assessment scheme is constructed it should be used for formative as well as summative evaluation. The aim of developing process aspects of mathematics is more likely to be achieved if pupils know of the aim; if they know of the specific process objectives; and if their work has regularly been assessed in terms of the achievement of these objectives.

If we want to be open in our mathematics teaching we can discuss the process objectives we have chosen and the grounds for their selection and ask our students to evaluate them.

Figure 5. An Example of a Short Investigation by a Pupil
Source: G.T. Wain & D. Woodrow, *Mathematics Teacher Education Project: Students' Material*, Glasgow, Blackie and Sons, 1980.

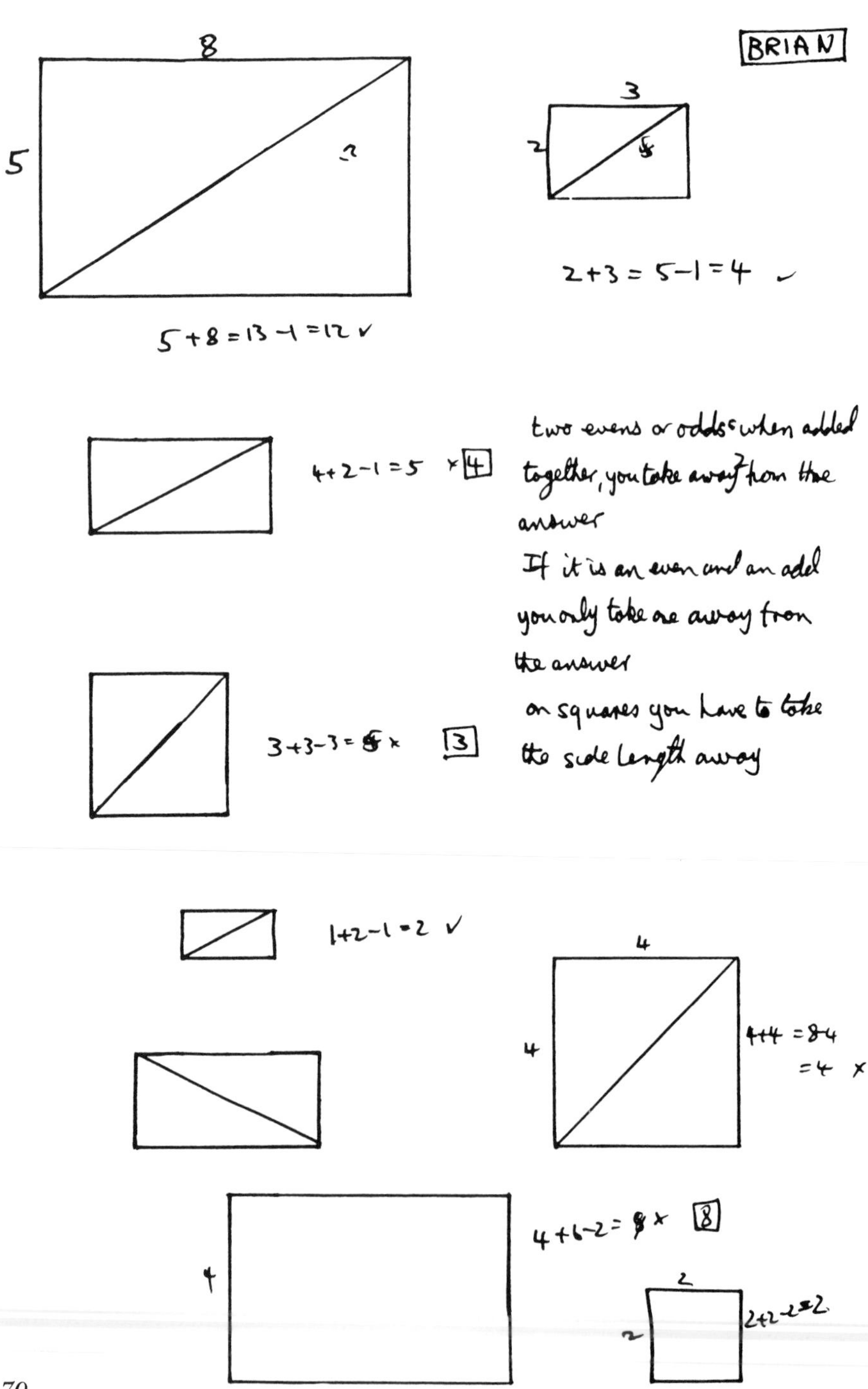

Conclusion

Much of this chapter has been devoted to discussing and proposing innovatory forms of assessment in mathematics, for work in the oral and the project or investigation modes. It seems appropriate to conclude by moving from a theoretical discussion to a concrete example from the classroom. Figure 5 illustrates the result of an investigation carried out by an above average fourth year secondary student, Brian. The problem investigated concerns the number of unit squares cut by the diagonal of a rectangular lattice (in the process of reproduction the faint background square grid has been erased). The investigation displays some crucial mathematical skills. Brian has been boldly and actively generating conjectures. But he has not systematically generated and organized his data; and his presentation skills, like many students', leave much to be desired. How should we teach Brian, not to mention average and below average students, to rise to the challenge of investigations?

Notes

1 K. Hart, *Children's Understanding of Mathematics 11–16*, London, John Murray, 1981.
2 Assessment of Performance Unit, *Mathematical Development*, Secondary Survey Report, No. 1, London, HMSO, 1980; *Mathematical Development*, Secondary Survey Report, No. 2, London, HMSO, 1981; *Mathematical Development*, Secondary Survey Report, No. 3, London, HMSO, 1982.
3 T.P. Carpenter *et al.*, *Results from the Second Mathematics Assessment of the National Assessment of Educational Progress*, Reston, Va., National Council of Teachers of Mathematics, 1981.
4 Royal Society of Arts, *The SLAPONS Scheme*, London, Royal Society of Arts, 1981.
5 B.S. Bloom, (Ed.), *Taxonomy of Educational Objectives: Cognitive Domain*, New York, David McKay, 1956.
6 Carpenter *et al.*, *op. cit*.
7 P. Ernest, *Educational Studies in Mathematics*, *15*, 1984, pp. 397–412.
8 H. Osborn, *Teaching Maths and Its Applics*, 2, 1983, pp. 36–42.
9 Assessment of Performance Unit, 1982, *op. cit*.
10 Mathematical Association, *Why, What and How*?, Leicester, Mathematical Association, 1976.
11 Department of Education and Science, *General Certificate of Secondary Education: The National Criteria for Mathematics*, London, HMSO, 1985 (reprinted in *Teaching Maths and Its Applics*, 4, 1985, pp. 1–6).
12 W.H. Cockcroft, *Mathematics Counts*, London, HMSO, 1982.
13 R. Sumner, *Tests of Attainment of Mathematics in Schools*, Windsor, NFER, 1975; I. Kyles and R. Sumner, *Tests of Attainment in Mathematics in Schools*, Windsor, NFER, 1977; Assessment of Performance Unit, *op. cit*., Note 2.
14 Assessment of Performance Unit, 1980, *op. cit*.
15 Department of Education and Science, *op. cit*.
16 P. Ernest, *Teaching Maths and Its Applics*, 3, 1984, pp. 80–6; Association of

TEACHERS OF MATHEMATICS, *Mathematics for Sixth Formers*, Nelson, Lancs, ATM, 1978; MATHEMATICAL ASSOCIATION, *Pupils' Projects, Their Use in Secondary School Mathematics*, Leicester, Mathematical Association, 1980; A.L. Hess, *Mathematics Project Handbook*, Reston, Va, NCTM, 1982.

17 MATHEMATICAL ASSOCIATION, 1980, *op. cit.*

18 ASSOCIATION OF TEACHERS OF MATHEMATICS, *op. cit.*

19 ERNEST, 1984, *op. cit.*

20 A. BELL, C. SHIU and B. HORTON, *Evaluating Attainment in Process Aspects of Mathematics*, Nottingham, Shell Centre for Mathematical Education, University of Nottingham, 1981.

21 K. HEDGER, and D. KENT, 'The Assessment of SIGMA Material,' Paper at the SIGMA Concerence, University of Exeter, 15–17 April 1985.

6 *Investigations: Where to Now?*

Stephen Lerman

Recently I observed a lesson in which an experienced mathematics teacher was asked by his Head of Department to 'do an investigation'. He chose one from a mathematical journal, and prepared it carefully beforehand, including writing out the completed solution on a piece of paper, which he held in his hand during the lesson. At the start he introduced the problem to the pupils, who clearly had never worked in this way before, and set them the task. He then went around the class, offering advice such as, 'no, not that way, it won't lead anywhere, try this', and 'that's right, keep on that way and you'll get the right answer' or 'don't give up, look here's the answer, on my paper'. It wasn't long before a pupil at the back put up her hand, and as if speaking for the whole class, asked the teacher to 'tell us how to do it now please sir'. The teacher managed to resist the request, and the pupils worked on. However, at the end of the lesson, the teacher commented to me that there didn't seem to be much in the 'business of investigations that was any different from normal mathematics lessons'.

There had been no opportunity for the teacher to discuss or examine what an investigation might be, how it might differ from 'normal' mathematics, how to conduct such a lesson, or any other aspects of problem-solving and investigations. Inevitably the investigative work of the class became just like any other lesson in mathematics, except that in this case the teacher seemed to the pupils to be behaving somewhat perversely, refusing to tell them how to do the problem before it was set or even during the lesson. It may also indicate, though this was just one school, how little contact there is between the work going on in most schools and the research or new developments taking place in the mathematical associations, in departments of education or amongst some teachers.

Our knowledge and experience of problem-solving and investigations has grown considerably in the last few years. Recently articles have appeared that look at the nature of this area of teaching mathematics, and what part it plays in mathematics education. It is my view that experiences like the above reveal the necessity of such fundamental considerations.

In this chapter, I intend to dig deeply into these issues, and to propose a theoretical framework to help structure our thinking and research in this area. I will examine first how problem-solving might relate to different conceptions of the nature of mathematics, and the issue of 'process versus content', and then support and develop the notion of Stephen Brown (1984) and others of a shift of discussion from 'problem-solving' to 'problem-posing'. Finally, I will suggest an alternative direction for development in the teaching of mathematics that differs radically from the direction in which we seem to be heading at the moment.

'Process versus Content' or 'the Chicken and the Egg'

The idea that in mathematics education we should be concerned with enabling pupils to acquire the skills and techniques that we identify as doing mathematics, or thinking mathematically, has been aired and discussed for some time. We always insist, though, that we must have content as well as a focus on processes. Since 'content' has traditionally been *the only* conscious focus of school mathematics, we seem to resort to specifying what we are going to teach, and leave the teaching of processes, or the development of problem-solving skills in pupils, in general vague terms, and as general guiding principles. The *National Criteria* (DES, 1985) exhibit this tendency. This is after all what we are good at, discussing whether calculus should be in the O-level syllabus, or linear programming, or geometry proofs.

There is no doubt that concentrating attention on processes is very problematic. For instance, we do not have any clear idea of how to teach 'generalizing' or 'reflecting', nor do we know whether these processes are carried over from one problem to the next. How does one assess understanding of 'conjecturing'? We seem to believe that we know how to assess understanding of content, and that is usually by a suitable examination question.

The major obstacle to a serious consideration of a process orientated mathematics curriculum may be our own reluctance, as teachers of mathematics, whose own mathematical education was so totally concerned with content, to think in terms of the processes of doing mathematics. As a personal illustration of this, I remember quite vividly my reaction several years ago to being offered the post of research mathematician in a team of scientists, constructing a mathematical model of the pollution of an important lake. I could not recall having studied 'Mathematical Models of Lakes' at university, and in a state of mild panic I turned the job down. I could not see myself as being able to *do* mathematics in any creative way, merely as being fairly able at remembering and reproducing what I had been taught.

My starting point is that there are many problems with 'content' as a focus, problems that challenge the teaching of mathematics in such significant ways that a fundamental reconsideration is necessary. First, as teachers of mathematics, we know only too well that a very small minority of pupils will be able to go on from such a school mathematics syllabus to use the skills or knowledge acquired in any situation other than school mathematics, i.e., not in other school subjects, in work situations or other adult-life needs. If there is any *minimum competency* demanded of school mathematics, it is surely the *adaptability* of knowledge and skills to changing problems, situations and needs in adult life and in employment (where there is any — an issue I will return to later). This is not a content issue. It is, of course, far from minimal in terms of the demands on mathematics education to 'provide' such competency!

A second problem with 'content' is that if we are certain that the mathematical knowledge we have is true absolutely, there is some justification in the area of content focused mathematics curriculum. The alternative epistemological position, that mathematical knowledge is fallible, has significant implications for mathematics education. First, let us briefly examine the issue of absoluteness and fallibility in the nature of mathematical knowledge. (For a more complete development of this, and the consequences for the teaching of mathematics, see Lerman, 1986). Since the 1930s we have been aware the Euclidean Geometry is simply a geometry, one of the possible ways of describing the physical structure of the universe, and that the decision on which geometry to use is thus not known a priori. The independence of the continuum hypothesis has done the same for arithmetic. Russell demonstrated the possible consequences of 'impredicative definitions' in mathematics, but the mathematics we use is full of them. There might be any number of barbers who both shave themselves and do not shave themselves! Lakatos has shown that the image we have of mathematics as proceeding by infallible deductions, from intuitively obvious assumptions to certain conclusions, is not correct. Mathematics, instead, develops by the re-transmission of falsity. Even the notion of truth is, it may be maintained, a relative one in mathematics, to be defined at the start of any discussion involving its use. (For a development of this thesis see Grabiner, 1974). We cannot expect any correspondence between the predictions of a mathematical theory and the objective facts of the outside world to confirm the truth of the theory, or determine which of rival theories is the correct one. What constitutes such a critical experiment, how the experiment is set up, and how the event is interpreted are all theory-laden decisions. The *certainty* we have is limited to the tautological, for example, if $3x + 7 = 2x + 11$ then $x = 4$, and even this, in theory at least, may lead to a contradiction. All this ought to lead us, as teachers of mathematics, to question the way we teach. If we, the possessors of this esoteric and certain knowledge, whose job it is to convey or reveal this certainty to pupils, are deluding ourselves, and that what characterizes mathematical knowledge is not its absoluteness but the particular ways of dealing with certain aspects of the world and society around us, then our status in relation to knowledge, and hence to pupils in the school situation, must change.

A third problem is what constitutes indispensable, irreducible, absolutely basic content? We cannot get beyond the Four Rules without confronting controversial issues such as the teaching of long multiplication, long division, directed numbers, number bases, etc., let alone the rest. We use all kinds of rationales for such decisions, such as statistically determined levels of competency for particular ages and topics (Hart *et al.*, 1981). As soon as one begins to question what must be included in a school mathematics syllabus, one falls back on what *content* will provide the *skills* that we all broadly identify as necessary in adult life, working or otherwise. As suggested above, whether content leads to skills for the great majority is doubtful.

Awareness of the central significance of context and meaning in mathematics education, as described recently by Cobb (1986), throws into question most of our mathematics teaching. We all know of instances of children with considerable skills outside the classroom that we would consider mathematical, who are unable to exhibit those abilities and confidence inside the classroom. The 'constructivist' view of children's learning focuses on what the child perceives in any situation, including what we as teachers present, and suggests that this necessarily has to be our starting point for understanding or interpreting what the child knows. It is inadequate to persevere with the idea that what matters is how we present the mathematical knowledge, how we explain it, and that doing this better will directly facilitate the child's learning. We have no hope of engaging in these issues as long as we hold on to 'content' as our focus.

If focusing primarily on 'content' is hard to justify, inadequate and unsuccessful, we should free ourselves from the constraints of our preconceptions, and examine the alternative. In the next section I shall look at what might be the consequences of reversing the order, and thinking of 'process' as coming before 'content'.

Mathematics through Problem-Posing

The assumptions for this part of the discussion are as follows. Since the context and the meaning are relative, and hence the engagement of the child is an individual response and not necessarily consequent upon the stimulus of the teacher, then offering the child an open situation, in which the child is encouraged to pose questions for her/himself, is the only way of enabling the child to advance conceptually. In a sense the child is doing this when s/he is learning, that is, providing the meaning and context that are 'meaningful' to that child. The problem is that most of what we do in the mathematics classroom is 'meaningless' to most children, and therefore is not learnt. Cobb (1986) describes the situation as follows:

> Self-generated mathematics is essentially individualistic. It is constructed either by a single child or a small group of children as they attempt to achieve particular goals. It is, in a sense, anarchistic mathematics. In constrast, academic mathematics embodies solutions to problems that arose in the history of the culture. Consequently, the young child has to learn to play the academic mathematics game when he or she is introduced to standard formalisms, typically in first grade. Unless the child intuitively realizes that standard formalisms are an agreed-upon means of expressing and communicating mathematical thought they can only be construed as arbitrary dictates of an authority. Academic mathematics is then totalitarian mathematics.

Mathematical knowledge de-reified, seen as a social intervention, its truths, notions of proof, etc. relative to time and place, has to be seen as integrally involved with the doing of mathematics, and indeed cannot be separated from it. Mathematics is identified by the particular ways of thinking, conjecturing, searching for informal and formal contradictions, etc., not by the specific 'content'.

An example of a problem that may best be termed a 'situation', in that asking the question is left up to the student, provides an illustration for the discussion that follows. In a seminar with a group of postgraduate students, who were non-mathematicians, I presented the following investigation, taken from SMILE *Investigations* (1981): 'Consider triangles with integer sides. There are 3 triangles with perimeter 12 units. Investigate' (see Figure 1).

A number of groups were disconcerted, saying that they had no idea what to do, since there was no question being asked. Other groups worked on areas of the triangles, perimeters of the triangles, perimeters of rectangles, etc. All the students found this to be quite different from the other investigations they had tried before, and very challenging. Brown (1984) describes a number of other examples like this one, where students de-pose or re-frame the question as stated, and generate their own problems.

It can be suggested that there are more radical consequences in changing from a 'content' to a 'process' focus. There are political implications of the notion of problem-posing, as suggested by Cobb's use of the terms 'anarchistic' and 'totalitarian'. Freire (1972), in writing about his literacy work with the oppressed people of Brazil and elsewhere in South America, describes two rival conceptions of education. The traditional view of education is the 'banking concept', whereby pupils are seen as

Figure 1. Diagram Accompanying Investigation
Source: SMILE, *Investigations*, London, Smile Centre, 1981.

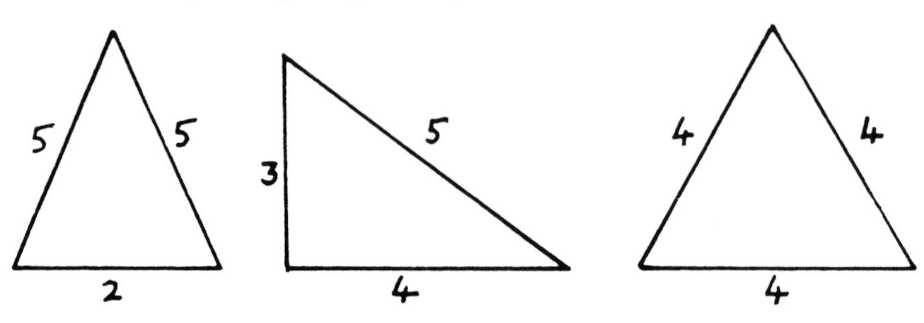

initially empty depositories, and the role of the teacher is to make the deposits. Thus the actions available to pupils are storing, filing, retrieving, etc. In this way, though, pupils are cut off from creativity, transformation, action and hence knowledge. The alternative view of education Freire describes as the 'problem-posing' concept. By this view knowledge is seen as coming about through the interaction of the individual with the world. 'Problem-posing' education responds to the essential features of the conscious person, intentionality and meta-cognition. Freire's discussion of opposing concepts of education is integrally tied with oppression and freedom.

We are working in education at a time when our students may be faced with a lifetime of unemployment and uncertainty; with threats to the ecology of the planet; an increase in disparity of wealth between rich and poor in society and between nations; and even the threat of total annihilation. Traditionally these have not been issues that have been thought of as having any relationship to mathematics. We have always rested safe in the knowledge that mathematics is value-free, non-political, objective and infallible. Freire's analysis, together with the doubts about such conceptions of the nature of mathematics and children's learning described above, suggest that this is not the case. Quite the contrary, we have a particular responsibility, since so many political, moral and other issues are decided using 'mathematical' techniques, to enable pupils to examine situations, make conjectures, pose problems, make deductions, draw conclusions, reflect on results, etc. These situations can be as in the investigation above, or the Fibonacci sequence (see Brown, 1984), information about expenditure on arms by the United States and the USSR, expenditure on education for different racial groups in South Africa, etc. There is plenty of 'content'. The last two situations would call on statistical and graphical techniques, for example. The difference is that engagement in the problems posed by the pupils puts mathematical knowledge in a different, and in my view appropriate, position. It is seen as a library of accumulated experience, and just as any library is useless to someone who cannot read, so too this library is useless unless people have access to it. When a problem is generated which reveals the need for some of this knowledge, be it multiplying decimals, standard index form, complex numbers, or castastrophe theory, if the individual recognizes first that such help is needed, and secondly that it is available, the context, relevance and meaning of mathematical knowledge are established.

Enabling students to examine situations and to pose problems for themselves reflects the fallibilist or relativist view of mathematical knowledge, reflects the constructivist perspective of children's learning, and places a powerful tool in the hands of people to examine what is happening to their lives, and provide them with the possibility of changing it.

Conclusion

Course work, extended pieces of work and investigations are becoming a compulsory part of the teaching of mathematics through their inclusion in GCSE mathematics. Far from these elements fundamentally changing the teaching of mathematics, we are in danger of losing what open-endedness exists in the mathematics classroom to examination-led criteria for assessment of children's work. Far from focusing on situations that relate to the context and meaning that children bring into the classroom, we are in danger of extending and reinforcing our judgments of childrens' 'abilities' in mathematics, based on school mathematics. (The use of quotation marks around the word 'abilities' is intended to devalue the notion. Given our growing awareness of some of the issues discussed here, we have to be highly suspicious of our usual use of the term.)

Perhaps the *National Criteria* should consist of processes of mathematical thinking, with some illustrations of the types of content areas that may be called upon by particular examples of situations that can be presented to pupils. Perhaps we should forget assessment for some years, until we have developed new ways of working in the classroom, and give every child a 'pass' in the meantime. Perhaps we should move towards a totally school-based assessment, as with English. Perhaps the criterion of success in a mathematics course should be the ability to take a newspaper, be it *The Sun* or *The Times*, or a piece of government legislation, and reveal some of the underlying assumptions and methods of deduction that have been used to reach conclusions and determine policies and attitudes that at the moment dominate our lives, and against which we feel ourselves to be powerless.

References

Brown, S.I. (1984) 'The Logic of Problem Generation: From Morality and Solving to De-Posing and Rebellion', *For The Learning of Mathematics*, 4, 1, pp. 9–20.

Cobb, P. (1986) 'Contexts, Goals, Beliefs and Learning Mathematics', *For the Learning of Mathematics*, 6, 2, pp. 2–9.
Department of Education and Science (1985) *GCSE, The National Criteria for Mathematics*, London, HMSO.
Freire, P. (1972) *Pedagogy of the Oppressed*, London, Sheed and Ward, London.
Grabiner, J.V. (1974) 'Is Mathematical Truth Time-Dependent?' *American Mathematical Monthly*, 81, pp. 354–65.
Hart, K. *et al.* (1981) *Children's Understanding of Mathematics: 11–16*, London, John Murray.
Lerman, S. (1986) *Alternative Views of the Nature of Mathematics and Their Possible Influence on the Teaching of Mathematics*, Unpublished PhD dissertation, King's College, University of London.
Smile (1981) *Investigations*, London, Smile Centre.

NEW CURRICULA

The 1960s were the golden years of curriculum development in Great Britain, with perhaps a score of new mathematics projects appearing, mostly in textbook or teacher's guide form. This activity quietened down in the early 1970s, which saw the development of fewer, but largely individualized schemes, often commercially developed. In primary schools this led to what has been perhaps the greatest change in the teaching of mathematics in thirty years — other than the broadening of content from arithmetic to mathematics. This was the shift from text-based whole class teaching to individual study from published schemes. The mid-to-late 1970s saw some notable exceptions to the overall pattern, such as the South Notts Project 'Journey into Maths', which pioneered a process orientated approach to secondary mathematics.

In the late 1980s there have been several new curriculum developments, funded from public or other non-commercial sources. Two of the most distinctive projects are 'Enterprising Mathematics' — a wholly integrated and contextualized secondary mathematics projects — reported in this section, and Nuffield Secondary Mathematics — based on research into the learning and teaching of mathematics, and directed by the distinguished researcher, Kath Hart. There have also been exciting teacher-led developments, such as the work in school associated with the PRIME Project, directed by Hilary Shuard, and the LAMP (1987) Project, directed by Afzal Ahmed.

Thus as we move into the 1990s there is a new wave of curriculum development in mathematics education. This is just as well, in view of the innovations reported above. For the new technology, the new assessment systems and new teaching approaches themselves represent a curriculum shift, and need support from newly developed textual materials. The mathematics curriculum for the 1990s needs to be substantially different from much of what has gone on in the past. The chapter that follows takes a more radical line, and questions the utility of mathematics, at least in the form that it is taught now. Whatever our views on this issue, it seems uncontroversial to say that much of the mathematics curriculum stands in need of reform. Whether the new developments will deliver what is needed, only time can tell.

Reference

LOW ATTAINERS IN MATHEMATICS PROJECT (1987) *Better Mathematics: A Curriculum Development Study*, London, HMSO.

7 *Mathematics Education for the Twenty-First Century: It's Time for a Revolution!*

David Burghes

Initiatives in Mathematics Education

There is so much activity in mathematics education post-Cockcroft that it is difficult to know where to begin. We have:

300 or so maths missionaries spreading the gospel throughout the land;
DES priority for mathematics in-service courses throughout the country;
new maths education centres at universities;
mathematics education research projects (e.g., slow learners, graded tests, investigations, etc.);
countless new books, GCSE schemes, packs, software, etc.;
new forms of assessment, eventually including grade-related criteria.

But will all this activity make any difference to mathematics education? There are two important questions to ask:

1 Will we produce a more numerate and adaptable generation?
2 Will we produce more school leavers who have enjoyed their maths at school?

Although most maths educators seem convinced that they have seen the light (e.g., investigations, practical work, discussion and problem-solving), I think it is time to question this new faith. I do not do this in order to put the clock back or to undermine confidence, but in order to provoke a *real revolution* in the way we teach (and pupils learn) mathematics.

The Faith

Almost everyone reacted favourably to the Cockcroft Report; it was particularly welcomed by those of us in the mathematics education community who for so long have been trying to show prospective teachers that there is more to a mathematics education than simply getting sums right. Yet mathematics at school and at university has continued to promote this view of mathematics. A well trained and hard working parrot could get a first class degree in mathematics at many universities — provided it could write rather than talk!

The Cockcroft Report (*Mathematics Counts*, published by the DES in 1982) has gained an international acclaim. Amongst many recommendations it advocates that *discussion, practical work, investigations* and *problem-solving* should be part of a mathematical diet as well as the usual theory and practice. So the faithful up and down the country now think that they have the answer. But just imagine bottom set fourth year late on a Friday afternoon: will it really motivate them to find out how many squares you can make on an $n \times n$ pegboard? Indeed, it probably will motivate them, if it is taught by a teacher with complete conviction and who has the personality to put it across in a meaningful way. But what about the average harassed, hard-pressed teachers — how are they going to cope?

It is not that I am against introducing such teaching methods into mathematics; I very positively encourage it, and with *good* teachers and *quality* in-service work I am convinced that these changes should take place. If nothing else, it makes teaching mathematics a far more unpredictable occupation. What I am not so convinced about is whether having made these changes (which will be both costly to implement and difficult for many teachers to cope with), we will have had much effect on giving positive answers to my questions (1) and (2) above. If the changes are made in a haphazard way with insufficient backing (and despite all the money that has been pumped into maths education, there is already evidence of this with the new GCSE courses), we might well have worsened the situation.

It is very easy to criticize, but not so easy to suggest alternative strategies, but I will at least attempt to put forward some tentative ideas. My suggestions (unlike Cockcroft's) might well be unpopular with the maths education community, but they are an honest attempt to suggest changes — no, more than changes — a *revolution*, which might prove more successful in the long run than our current attempts at change.

An Alternative Faith: Motivation

Pupils will work at mathematics (or any other subject) if they are motivated to do so, so how do we provide the motivation for mathematics? It will depend on such factors as:

the personality and talent of the teacher,
the school environment, and class associates,
the home environment,
job prospects (or lack of),
the curriculum,
assessment,

and many other things. I put the teacher first as I believe that he or she is the most crucial motivator — just think back to your own time at school. Although children have particular talents for various topics, a good teacher can make all the difference between a pupil losing interest (which is exceedingly difficult to recover in mathematics) or working conscientiously, enjoying the subject. So my faith starts with the teacher: if somehow we could take the most talented dozen or so maths teachers in the country and make thousands of clones, many of our problems in maths education would be solved. But, and this is where I am on very shaky ground, all my experience shows that good maths teachers are born with the talent, and I am not yet convinced that training does much good. I liken this to my experience as a potential Middlesex cricketer! At the age of 14 I was 'spotted' by my school after a spectacular five-wicket fast bowling spell in a house game (which was the first time I had bowled in a competition match); during the winter I was sent to the MCC for training. I know that the coaches were doing their best, but I spent three further years representing the school at cricket, and I never took more than two wickets in one innings again! My inelegant style and jumping action had been replaced by a nice, clean, smooth action but my bowling was doomed for evermore. Gone was the enthusiasm and confidence, replaced by a professionalism that was not needed! Are we in danger of doing the same when we train maths teachers?

Although school, home and job prospects come next on my list, they are outside the scope of this chapter and so I look next at the curriculum and its assessment. On the curriculum we can perhaps make some progress, as certain topics are more motivating than others. In Appendix 1 we show an algorithm for finding the day of the week for dates this century — you can use this for classes from age 11 up to 100 — which always motivates, not because pupils are keen to know why it works, but because *they want to know the answer*. They want, for example, to know which day of the week they were born on, and they enjoy performing the calculation to get the answer — the computer could do it all for you, but that takes the fun out of the problem. So a key seems to be in finding activities in which pupils need to use mathematics to find out results that *they* want to find out. It is probably not possible to rewrite the whole maths curriculum in such a way that pupils are always motivated in this way, but we all know that too often pupils are performing calculations or learning techniques that have absolutely no relevance for them at all.

So my faith begins with *motivation* through

1 the ability of good teachers, and
2 the relevance of the curriculum.

So far this faith is compatible with the post-Cockcroft faith, but my interpretation will start to diverge from others.

Relevance of the Curriculum

I will start by assuming that for most people there is little or nothing in GCSE mathematics beyond List 1 (see Appendix 2) that is directly relevant for future life. For those continuing onto A-level and into higher education there is a natural progression upwards, but for the vast majority of children who complete their mathematical instruction at age 16 there is precious little in secondary mathematics that we can in all honesty defend as being directly useful for future life. That is not to say that there are not good reasons for the inclusion of many topics, but direct relevance or usefulness is certainly not the only reason. Topics of direct relevance, such as basic numeracy, percentages, weighing and measuring, money transactions, reading timetables and betting odds do not constitute a large part of the mathematics curriculum, many are dealt with at primary rather than secondary level.

So with what should we fill the maths curriculum? One suggestion is very simple: don't fill it up at all! Why do we continue to protect mathematics as a subject that needs two to three hours every week for five years in secondary schools? Our usual meaning of mathematics probably needs such time — and probably much more for many pupils. But why bother at all? Why put our children through this misery week after week? Surely we can think of something better and more positive to do with the time. Could we accept the thesis that children do not have to succeed at *academic* mathematics to be worthwhile citizens? What we must aim for is to make sure that they are numerate (and this often means making sure that they do not get worse at sums when they leave secondary school than when they entered!) and that they are offered something mathematical which is enjoyable, creative and stimulating, whilst enabling them to reach their individual mathematical potential.

Another suggestion is to teach mathematics through applications and contextual situations. This is the theme of the 'Enterprising Mathematics' project, which is a modular approach to teaching GCSE mathematics. Here, at least, pupils will see the relevance of mathematics. The modules are in business and commerce, design, science and technology, environment and leisure and recreation. So in some way, we have taken mathematics off the curriculum. (Full details of this project are given by David Hobbs in Chapter 8.)

Challenge in Mathematics

I have already outlined the theme of motivation through relevance, but relevance is not enough. Challenge is also needed, but the appropriate challenge for each individual. Appendix 3 gives an example of a worksheet on the classical 'Travelling Salesman Problem'. It is structured so that all students can make progress: they can all find a possible solution, and the challenge is to find the best possible solution. In Appendix 4 there is another example of a problem which has a steady build up through to the final problem (which is not as easy as it looks). Challenge is, like relevance, an important ingredient that provides the motivation needed for effective and enjoyable teaching and learning of mathematics.

New Technology

Elsewhere in this book you will learn more about new technology and its possible impact for teaching and learning of mathematics. At this stage it is very easy to become obsessed by interactive video (its potential is understandably enormous; its acceptability as a teaching medium still remains to be proved), but we must not forget the potential of earlier technologies (audio-tapes, video and computer software). We should also note that we already have hand-held calculators which can give a graphic display for functions, and it will not be long before we have small calculators capable of formal algebra and calculus; i.e. given a function, say x^2, they will, at the press of a button, give the answer '$x^3/3 + c$'. The existence of such programs (currently available for mainframe computers, although packages are now being developed for microcomputers) should initiate a debate over the content of school (and university) mathematics curricula. Indeed, it gives strength to the argument that it really does not matter what mathematical content we teach since it can all be done by computers now — it is the process skills that are important. These are skills such as:

understanding problems;
implementing strategies for solution;
interpreting data;
transferring skills to new problems;
tackling unseen problems;
critical judgment;
working positively in a group.

It is this type of criterion which is far more important to later life than the usual stress on yet more and more abstract content.

Finally though, I would not like to see all of mathematics go out the window. Just doing mathematics can be an enjoyable activity for many people — it should be encouraged, since there are so many beautiful and elegant aspects to mathematics, for example: four-colour theorem for maps, perfect numbers, Fibonacci sequence, Golden Ratio, Mobius Strips. Unfortunately, topics of this sort do not occur on conventional syllabuses.

Final Remarks

I have tried to outline possible developments in the teaching and learning of mathematics over the next decade. Any historical perspective gives little cause for comfort — the Cockcroft Report is just the latest of a long series of reports into what is wrong with the teaching of mathematics that have been published over the past century. We have always had problems. Maybe the heart of the problem is that mathematics (as we conventionally see it) is a difficult subject, and there is no chance of any dramatic improvement (unless we move the goalposts!).

Surely the mathematics education community must recognize that currently we do much harm to many pupils at school. My rough estimate would be that more than half the population leave school with a distaste for the subject. They become the adults of tomorrow who cannot cope with even simple sums. Mathematics is a subject that can be *enjoyed* by everyone. But, although enjoyment in mathematics is my ultimate goal, it should also be stressed that the nation needs more able generations, who are adaptable, flexible and able to use their mathematical abilities. We need a revolution in the way we teach mathematics. Maybe I am alone in my thinking, but I cannot be convinced that the presently planned changes in mathematics teaching will produce anything but chaos. If we are going to have chaos, then let's go the whole way and have revolution — we might ultimately achieve more!

Appendix 1

Birthday Algorithm

Step 1

Find $S = Y + D + \left[\frac{Y - 1}{4}\right]$,

where Y = Year

D = Day of Year (Jan 1st = 1, ..., Jan 31st = 31, Feb 1st = 32, ...)

$\left[\frac{Y - 1}{4}\right]$ means ignore the remainder

Step 2

Divide 5 by 7 and note the remainder

Step 3

The day of the week is given by the key

Remainder	Day
0	Friday
1	Saturday
2	Sunday
3	Monday
4	Tuesday
5	Wednesday
6	Thursday

Example : For the birthday of the Princess of Wales, we have the birth date July 1st, 1961.

Step 1

Y = 1961

D = 31 + 28 + 31 + 30 + 31 + 30 = 1 = 182

$$\left[\frac{Y - 1}{4}\right] = \left[\frac{1960}{4}\right] = 490$$

So S = 1961 + 182 + 490 = 2633

Step 2

S divided by 7 gives 376 and remainder 1

Step 3

Using table, 1 corresponds to SATURDAY

So the Princess of Wales was born on a Saturday.

Appendix 2

Whole numbers: odd, even, prime, square. Factors, multiples, idea of square root. Directed numbers in practical situations. Vulgar and decimal fractions and percentages; equivalences between these forms in simple cases; conversion from vulgar to decimal fractions with the help of a calculator.

Estimation.

Approximation to obtain reasonable answers.

The four rules applied to whole numbers and decimal fractions.

Language and notation of simple vulgar fractions in appropriate contexts, including addition and subtraction of vulgar (and mixed) fractions with simple denominators.

Elementary ideas and notation of ratio. Percentage of a sum of money. Scales, including map scales. Elementary ideas and applications of direct and inverse proportion. Common measures of rate.

Efficient use of an electronic calculator; application of appropriate checks of accuracy.

Measures of weight, length, area, volume and capacity in current units. Time: 24 hour and 12 hour clock. Money, including the use of foreign currencies.

Personal and household finance, including hire purchase, interest, taxation, discount, loans, wages and salaries. Proft and loss, VAT. Reading of clocks and dials. Use of tables and charts. Mathematical language used in the media. Simple change of units including foreign currency. Average speed.

Cartesian coordinates. Interpretation and use of graphs in practical situations including travel graphs and conversion graphs. Drawing graphs from given data.

The use of letters for generalised numbers. Substitution of numbers for words and letters in formulae.

The geometrical terms: point, line, parallel, bearing, right angle, acute and obtuse angles, perpendicular, similarity.

Measurement of lines and angles. Angles at a point. Enlargement.

Vocabulary of triangles, quadrilaterals and circles; properties of these figures directly related to their symmetries. Angle properties of triangles and quadrilaterals.

Simple solid figures.

Use of drawing instruments. Reading and making of scale drawings. Perimeter and area of rectangle and triangle. Circumference of circle. Volume of cuboid.

Collection, classification and tabulation of statistical data. Reading, interpreting and drawing simple inferences from tables and statistical diagrams. Construction of bar charts and pictograms. Measures of average and the purposes for which they are used.

Probability involving only one event.

Appendix 3

TRAVELLING

A computer salesman, who lives in London, wants to visit all the cities shown on the map.

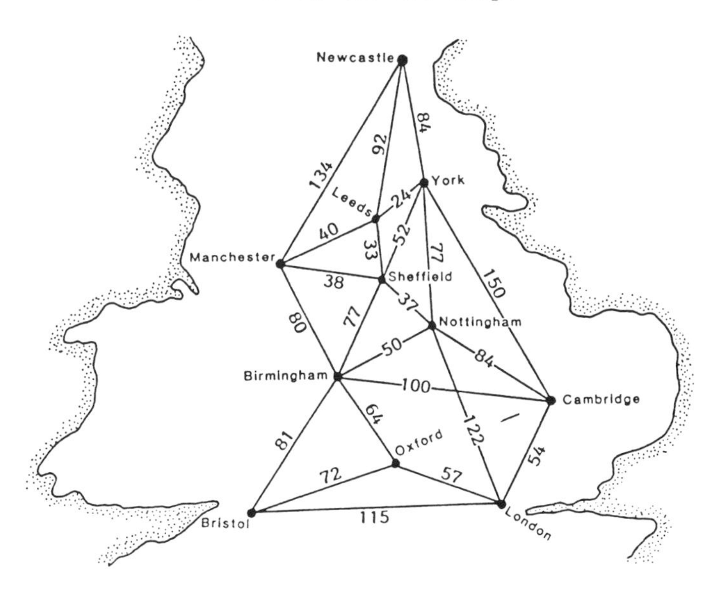

1. Find a route for him.
2. What is its total length?
3. Can you find a route of length less than 750 miles?
4. How long is the shortest route you can find?
5. If he no longer has to visit Sheffield, what is his shortest route?

An Edward Arnold Master

Appendix 4

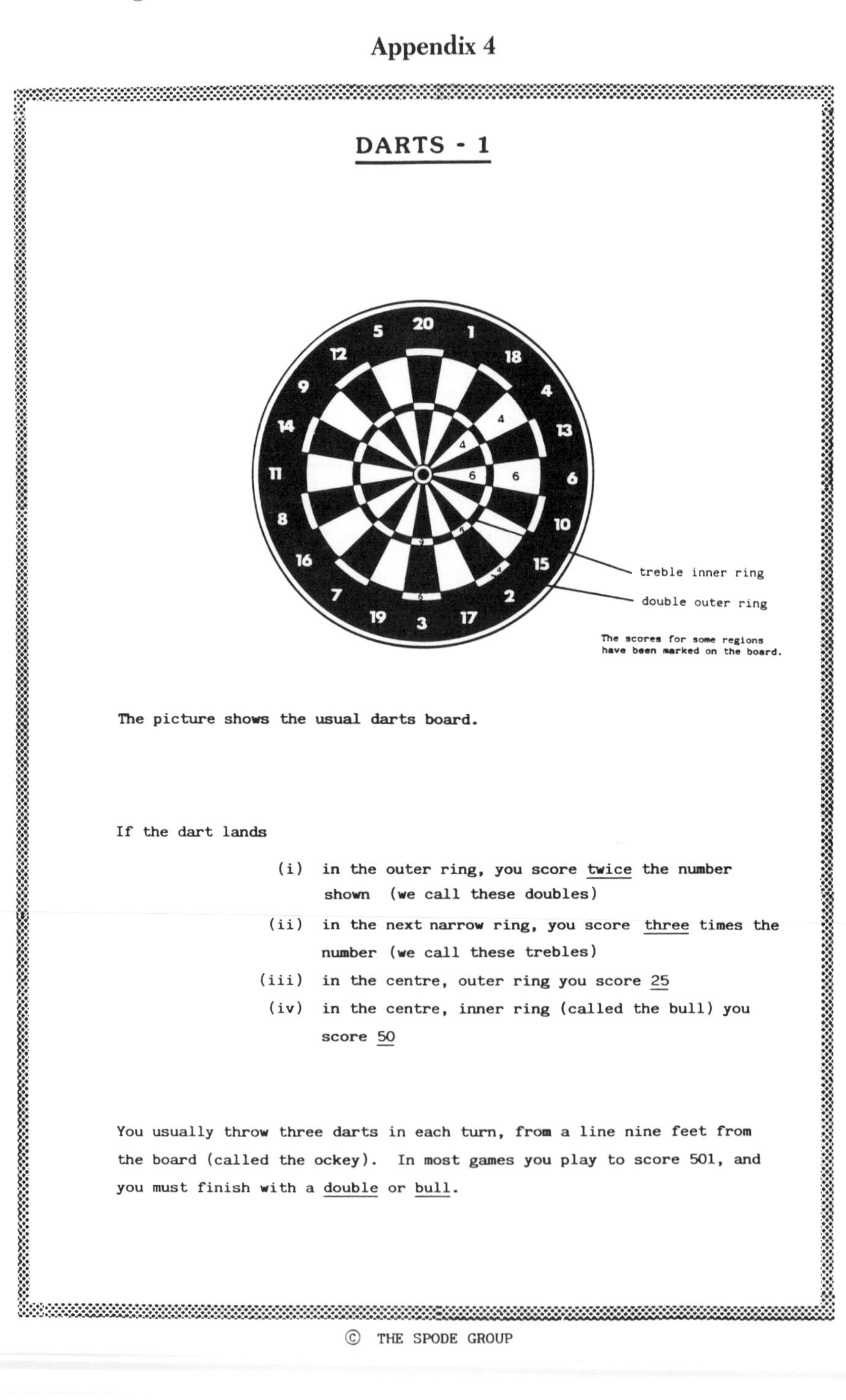

DARTS - 1

treble inner ring

double outer ring

The scores for some regions have been marked on the board.

The picture shows the usual darts board.

If the dart lands

(i) in the outer ring, you score twice the number shown (we call these doubles)

(ii) in the next narrow ring, you score three times the number (we call these trebles)

(iii) in the centre, outer ring you score 25

(iv) in the centre, inner ring (called the bull) you score 50

You usually throw three darts in each turn, from a line nine feet from the board (called the ockey). In most games you play to score 501, and you must finish with a double or bull.

DARTS - 2

PROBLEMS

1. Make a list of all the integers from 1 to 60. Indicate which of these numbers can be thrown with a single dart.
   Which of these numbers are doubles?
   Which of these numbers are trebles?

2. How many numbers between 1 and 60 cannot be thrown with one dart?

3. What is the most that you can score with 3 darts?

4. You need 57 to finish and it is your turn. With your first dart you score 8. How many ways are there of finishing, using the next two darts?
   (Remember – you must finish with a double or bull.)

5. Find the least number of throws needed to score 501 (ending with a double or bull).

6. Can you finish in three darts, if you need to score

   (a) 61 (b) 72 (c) 89 (d) 111 (e) 132
   (f) 145 (g) 166 (h) 170 (i) 177?

7. Experts aim for treble 20 (often scoring 20). If, however, you just miss 20, you would land in 1 or 5.
   Suppose you are not an expert dart thrower. Which sector should you aim for in order to score as much as possible?

8. What is the smallest number (not including 1) that you cannot score with at most three darts in order to finish?

8 *The 'Enterprising Mathematics' Project*

David Hobbs

In *Aspects of Secondary Education in England* (DES, 1979) attention was drawn to two aspects of mathematics teaching. First, the style: most lessons were of a stereotyped nature — the teacher presented a topic on the blackboard, worked through an example, and then the pupils did practice exercises; secondly, the content: the mathematics seldom related to anything else, and it was not inherently challenging. These points were followed up by the Cockcroft Report, *Mathematics Counts* (DES, 1982). The celebrated paragraph 243 suggested that mathematics teaching at all levels should include opportunities for:

exposition by the teacher;
discussion between teacher and pupils and between pupils themselves;
appropriate practical work;
consolidation and practice of fundamental skills and routines;
problem-solving, including the application of mathematics to everyday situations;
investigation work.

Later (paragraph 462) an extract from a submission is quoted:

> Mathematics lessons in secondary schools are very often not about anything. You collect like terms or learn the laws of indices, with no perception of why anyone needs to do such things. There is excessive preoccupation with a sequence of skills and quite inadequate opportunity to see the skills emerging from the solution of problems.

The GCSE *National Criteria for Mathematics* (DES, 1985) broke new ground by including objectives which could not be fully tested through written examinations:

> 3.16 respond orally to questions about mathematics, discuss mathematical ideas and carry out mental calculations;

3.17 carry out practical and investigational work, and undertake extended pieces of work.

The *National Criteria* also state that between 20 and 50 per cent of the overall assessment has to be given to course work. With regard to content the *National Criteria* are less adventurous. Dull lists of mathematical topics are provided for the least able group (List 1) and the middle ability group (List 2). For example, in List 2, we have

Basic arithmetic processes expressed algebraically.
Directed numbers.
Use of brackets and extraction of common factors.
Positive and negative integral indices.
Simple linear equations in one unknown.
Congruence.

A syllabus for the most able group is not provided in the *National Criteria*, the GCSE boards being allowed freedom to decide their own (although the contents of List 2 must be included). The GCSE boards have responded by producing helpful material and guidelines for course-work, but in most cases their syllabuses are Lists 1 and 2 of the *National Criteria* with a few words of explanation included.

Mathematics in Context

Most people believe, often naively, that mathematics is 'useful', by which they mean that it is of use in everyday life and in many jobs. This is a view which parents, employers and politicians have propagated and which has an influence on children. In the classroom teachers often use it as a carrot, although much of what is done — algebraic manipulation, congruence, etc. — seems far removed from being useful. The response to 'What is the point of this?' is often that the use will be perceived at a later stage, the argument being that we will do the mathematics now and you will apply it later. However, in practice the applications seldom emerge and for most pupils this later stage never comes.

An alternative approach, which is by no means new, is to recognize that much of mathematics has been developed in response to problems arising in the world around and to start with these contexts. Thus the everyday experience of sending a letter can be the starting point for a variety of activities. A discussion about how a letter gets from Manchester to London in a few hours can lead to points such as:

Where are local post boxes sited? (Are there regulations about distances between post boxes? In view of housing development should the local boxes be resited?)

Figure 1. A Sample Advertisement and What It Entails

If you make the minimum deposit of 20% the rate is 4.9% **(APR 9.5%)**. If you deposit 50% or more the rate is even less. Only 2.5% **(APR 4.8%)**.

Cash price	£5452.86
9.5% APR Initial Payment (Minimum 20%)	£1090.57
36 Monthly Payments of (Starting 1 month after contract)	£138.99
Charge for Credit	£641.35
Total Credit Price	£6094.21

How are collection and delivery routes planned? (elementary ideas of operational research)
How are post codes allocated? (leading to discussion about other coding system — ISBN, bar-codes)
How is mail sorted using post codes?
How efficient are first and second class posts?

Also at a low level it can involve practical work, first estimating weights of letters and parcels and then weighing them, and it can involve skills of obtaining information from tables in Post Office leaflets. The problem of designing a useful book of stamps to cost £1, say, can appeal at all levels.

A contextual approach is equally valid for able children. Interpretation of the rules for maximum dimensions of parcels soon leads to higher-level mathematics. An advertisement such as the one shown in Figure 1 raises the question of how the APR (annual percentage rate) has been calculated. This immediately leads to some interesting mathematics requiring the summation of a geometric sequence and the solution of an equation of degree 37. The use of a microcomputer together with trial and error techniques makes this accessible to able pupils. Such an approach uses the motivation of exploring contexts to which pupils can relate, and beginning where pupils are in their conception of what mathematics is about.

The Enterprising Mathematics Project

A curriculum development project to explore this contextual approach, 'Enterprising Mathematics', has been set up at the Centre for Innovation in Mathematics Teaching, Exeter University, funded initially by the Department of Trade and Industry. The aim is to produce material which can be used as a GCSE course for secondary years 4 and 5 and also as a resource in further education courses such as CPVE. Influenced by TVEI

Figure 2. The Structure of 'Enterprising Mathematics'

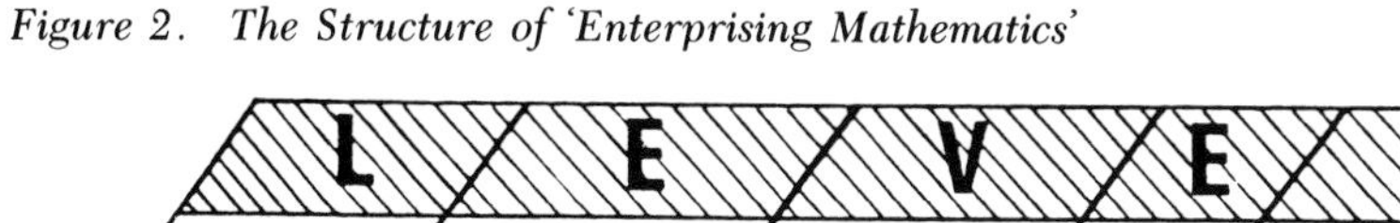

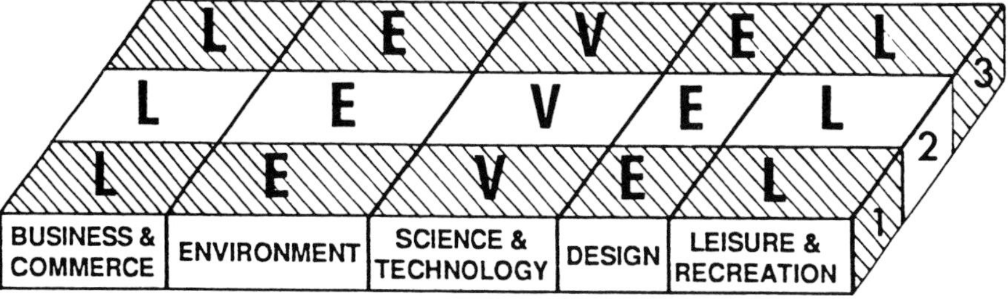

thinking, it is planned to organize the course on a modular basis in five blocks:

Business and commerce
Environment
Design
Science and technology
Leisure and recreation

Each block will contain material for one term's work in a GCSE mathematics course and will be available at the three levels referred to in the *National Criteria for Mathematics* (see Figure 2).

The main feature of the course is that it will develop mathematical concepts and techniques from real contexts. Practical work, applications, investigations and problem-solving will arise naturally from these contexts. Thus the realization of GCSE objectives 3.16 and 3.17 will be an essential part of the pupil's work and not a 'Friday afternoon' activity. Fifty per cent of the overall assessment will be given to course work. Using appropriate contexts the mathematical content of the *National Criteria* lists will be covered. There will be a need to consolidate and practise, and material for that purpose will be provided, but it should be seen as supplementary and not as the main component. The course will aim to show that mathematics is about something and that it relates to other school subjects. Collaborative teaching will be encouraged, thus making the material suitable for use in the curriculum structures being developed through TVEI and CPVE.

From of the Materials

Recent improvements in reprographic technology have made it easier for teachers to produce their own material, and many teachers are now looking for a more flexible resource than a single textbook. In particular, desktop publishing opens up possibilities for the production of material by a curriculum development project.

The trial version of 'Enterprising Mathematics' is being produced as photocopyable master sheets. Some topics have been written as booklets involving a progression of ideas with each page or double page being complete in itself. Thus teachers can put the pages together to make a structured booklet or use the pages singly. Either way there is the flexibility to insert other material or to replace pages. Some topics lend themselves to an unstructured format, and the material is presented as one-off sheets giving activities for groups or individuals. Thus whilst providing a course which is, as far as possible, complete, there is flexibility for teachers to personalize it.

A further advantage of providing photocopyable master sheets is that updating is easier than with a textbook. This is particularly important where contextual material is being used — prices of cars, interest rates, etc. can soon become out-of-date. Looking further ahead, the establishment of electronic databases such as NERIS will enable even more rapid updating to take place.

It could be argued that in order to encourage the styles of teaching recommended in paragraph 243 of the Cockcroft Report it would be better to produce material for teachers rather than for pupils — written pupil material can inhibit discussion and practical work. Although such an approach might be welcomed by adventurous, independent teachers, the evidence from the past suggests that such material is not used by most teachers. For example, the Schools Council 'Mathematics for the Majority' (1970–1974) project produced books for teachers, but their influence was slight, and a subsequent project was set up to produce pupil material. An intermediate approach, which is being explored, is to put the discussion ideas in the pupils' material, not for the pupils to work at alone, but for use as starting points by the teacher, as is illustrated in two sample pages (Figures 3 and 4). Figure 3 is a sample page from the Level 1 module on business and commerce; Figure 4 is from the Level 3 module in the same subject area. The difficulty is to keep written material fairly open, without closing it down, yet to give sufficient help for teachers and pupils to work on. For example, in a draft unit on Time (for the science and technology block) the intention is to incorporate practical work which involves making simple devices to measure time. An open approach is to pose the problem, 'You are shipwrecked on a desert island. Invent a method for measuring time.' In a draft version a stimulus page has been included consisting of pictures showing various historical methods (sand clock, water clock, candles, pendulum, etc.). Should there be further pages giving constructional details of such devices? Again, the problem of determining the day of the week for any given date arises. Should it be posed and then left entirely for discussion, should a structured approach be developed or should a procedure be given? To what extent should the amount of guidance depend on the ability of the pupils? These are the sorts of questions the project team is having to answer. For example, in

Figure 3. Sample Page from Level 1 Module on Business and Commerce

E: Sarah is also a saver!

Managing Money

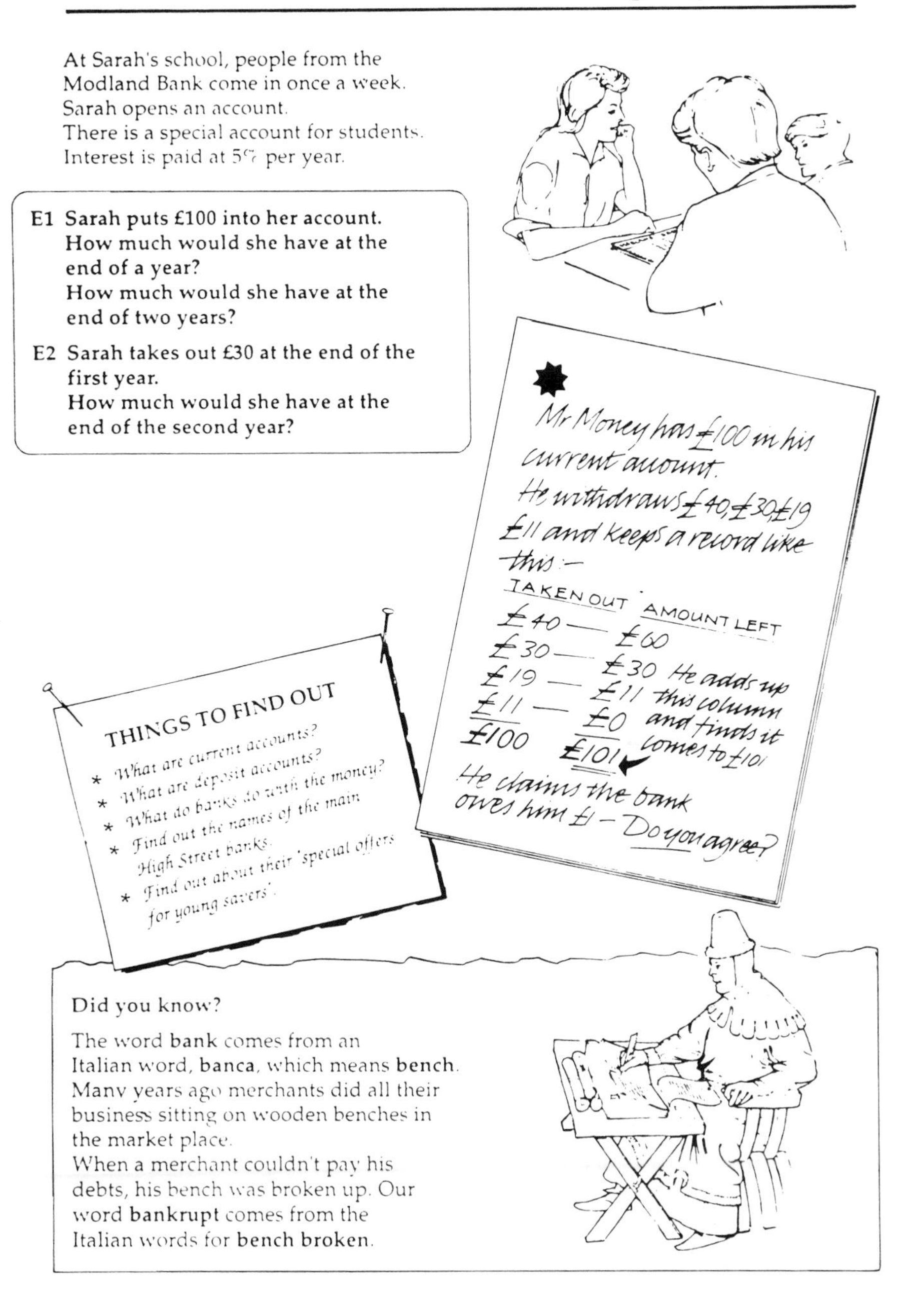

At Sarah's school, people from the Modland Bank come in once a week.
Sarah opens an account.
There is a special account for students.
Interest is paid at 5% per year.

E1 Sarah puts £100 into her account.
How much would she have at the end of a year?
How much would she have at the end of two years?

E2 Sarah takes out £30 at the end of the first year.
How much would she have at the end of the second year?

Mr Money has £100 in his current account.
He withdraws £40, £30, £19, £11 and keeps a record like this:—

TAKEN OUT	AMOUNT LEFT
£40	£60
£30	£30
£19	£11
£11	£0
£100	£101

He adds up this column and finds it comes to £101.
He claims the bank owes him £1 — Do you agree?

THINGS TO FIND OUT

* What are current accounts?
* What are deposit accounts?
* What do banks do with the money?
* Find out the names of the main High Street banks.
* Find out about their 'special offers' for young savers'.

Did you know?

The word **bank** comes from an Italian word, **banca**, which means **bench**.
Many years ago merchants did all their business sitting on wooden benches in the market place.
When a merchant couldn't pay his debts, his bench was broken up. Our word **bankrupt** comes from the Italian words for **bench broken**.

Figure 4. Sample Page from Level 3 Module on Business and Commerce

Saving

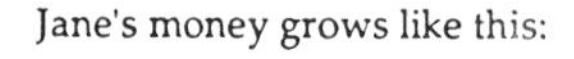

Jane's money grows like this:

Amount at the beginning of this year = **Amount at the beginning of last year** x **1.06**

There is a repetitive process here:

multiply by 1.06
multiply by 1.06
multiply by 1.06
etc.

Multiply by 1.06

```
10 INPUT "Amount "A
20 INPUT "Percentage "P
30 INPUT "Number of years "N
40 C=0
50 REPEAT
60 C=C+1
70 PRINT C,A
80 A=A*(1+P/100)
90 UNTIL C=N+1
```

When the interest rate is P% per year the multiplying factor is

$$1 + \frac{P}{100}$$

This multiplying factor can be seen in line 80 of the computer program.

Here is a printout for Jane's investment:

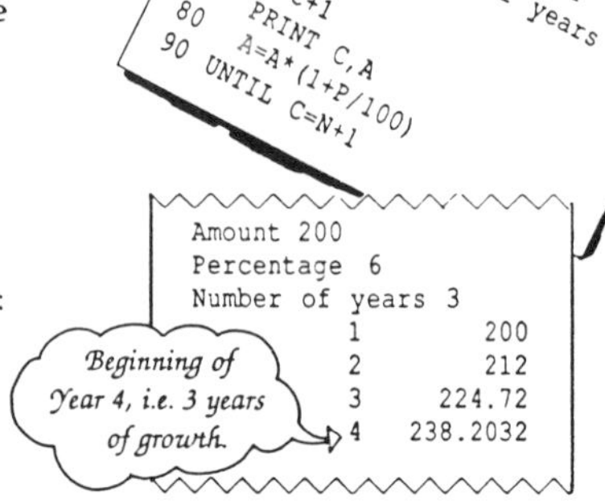

```
Amount 200
Percentage 6
Number of years 3
1       200
2       212
3       224.72
4       238.2032
```


A5 Try the program with the examples on page 2.

A6 One hundred years ago, Jane's great-great-great grandfather invested £1 at 5%. What is it worth now? More realistically, try it with a variable investment rate, for example, 3% for the first 20 years, 4% for the next 20 years, etc.

Follow-up activities:

addition to the topic 'booklets' the material contains backup sheets providing consolidation and practice together with 'gap-filling' to allow for the different courses pupils will have followed in previous years.

Teachers will be encouraged to use a variety of different styles by the range of supporting materials. It is planned to produce a newspaper containing articles of current interest, advertisements, data (finance, sport, etc.), puzzles, etc., presented in a lively style. It can be dipped into when appropriate by the teacher and used for stimulus activities. An advantage of a newspaper is that it can be updated cheaply without becoming fossilized like a textbook.

The use of microcomputers will be encouraged by provision of software consisting of simulations, investigations, for use by the teacher from 'out front' and by individual children, data files, and a version of some of the backup material. Audio-tapes, containing conversations in context, radio-style programmes, simulated discussions, etc., will be provided to exploit an approach used in other school subjects but not often in mathematics. Video has great potential for showing contexts outside the classroom. Material relevant to mathematics is already available through television and film libraries. Further videos will be produced to accompany the course. Interactive Video brings together the facilities of video disc and microcomputers, thus giving instant access to pictures and allowing decision-taking with feedback. A disc on the theme of running a school disco has been produced which will be a valuable resource in the business and commerce block.

'Enterprising Mathematics' aims to tackle the two issues of content and style identified in the reports referred to in the introduction. It is hoped that by providing motivation the course will enable students to enjoy their mathematics at school, reaching their mathematical potential and gaining confidence in their ability to use and apply mathematics in a variety of situations.

References

Cockcroft, W.H. (1982) *Mathematics Counts*, London, HMSO.

Department of Education and Science (1979) *Aspects of Secondary Education in England*, London, HMSO.

Department of Education and Science (1985) *GCSE: The National Criteria*, London, HMSO.

Schools Council (1970–1974) *Mathematics for the Majority*, London, Chatto and Windus Educational.

Part II

New Research Perspectives

In the 1980s there has been a professionalization of mathematics teaching through the growth of knowledge. Not only has mathematics education research increased — both in quantity and in quality — but so has teachers' awareness of it. Key publications have been important in this respect, especially the Cockcroft Report (1982), which referred to significant research results. Much of this concerned understanding and achievement in mathematics, explored both by large-scale testing and by interviews, and reported in APU (1979–1982, 1985), Hart (1981) and Sewell (1981).

The Cockcroft Committee commissioned a number of surveys of research by leading British workers in mathematics education. These surveys were used as authoritative sources and references in the report. They have since been published (see Bell, Costello and Kuchemann, 1983; Bishop and Nickson, 1983; Howson, 1983), and stand as significant contributions to the literature, and as vehicles for the dissemination of research results. Further research surveys have been published in the 1980s, especially Dickson, Brown and Gibson (1984) and Shuard (1986), both of which have been widely taken up by teachers (the latter is, in fact, more than a survey).

A major avenue for the dissemination of research findings has been the mathematics teaching associations: The Mathematical Association and the Association of Teachers of Mathematics, both through their journals *Mathematics in School* and *Mathematics Teaching* and through their annual conferences. In addition to dissemination, these organizations have encouraged teachers to carry out research, to view the teacher in the role of teacher-researcher. Of course, a key role in British research in mathematics education is also played by the higher education institutions. These offer a range of courses with research components including first degree work, Mathematical Association diplomas, Masters degrees and doctoral research programmes.

At the leading edge British research in mathematics education is presented at the conferences of national organizations such as the British Society for Research into the Learning of Mathematics, and at international conferences such as the International Congress of Mathematical Education (ICME) and those of the Psychology of Mathematics Education group (PME). However, not all research in mathematics education is carried out by those identified with the field. Mathematics has long been a favoured area of inquiry for educational researchers in general, especially educational psychologists. Important contributions to our knowledge of the teaching and learning of mathematics have been made by researchers such as Neville Bennett (Bennett *et al.*, 1984) and Charles Desforges (Desforges and Cockburn, 1987), amongst others. Both these researchers have contributed chapters to this part of the book.

As well as making contributions to the field, researchers from beyond the mathematics education community have voiced some criticism of the current research scene. It is claimed that much of the recent research on the teaching of mathematics is overly empirical, with insufficient theoretical underpinning. Another complaint is that it is steeped in a romantic ideology — pro-discussion, activity, investigation, etc. — and insufficiently self-critical. There is some substance to these claims. The remainder of this book — at least in part — tries to address these issues. This part of the book treats three areas: first, the close application of research perspectives to the teaching of mathematics; second, aspects of the romantic ideology of mathematics teaching from a critical research perspective; third, aspects of an important theoretical perspective on the learning of mathematics — the constructivist view of learning. A further growth point in research in mathematics education — the social context of mathematics teaching — is omitted, but only because it has the third part of the book to itself.

References

ASSESSMENT OF PERFORMANCE UNIT (1979–1982) *Mathematical Development, Primary Survey Reports 1–3, Secondary Survey Reports 1–3*, London, HMSO.

ASSESSMENT OF PERFORMANCE UNIT (1985) *A Review of Monitoring in Mathematics 1978–1982*, London, Department of Education and Science.

BELL, A.W., COSTELLO, J. and KUCHEMANN, D. (1983) *A Review of Research in Mathematical Education, Part A, Research on Learning and Teaching*, Windsor, NFER-Nelson.

BENNETT, S.N., DESFORGES, C., COCKBURN, A.D. and WILKINSON, B. (1984) *The Quality of Pupil Learning Experiences*, London, Erlbaum.

BISHOP, A.J. and NICKSON, M. (1983) *A Review of Research in Mathematical Education, Part B, Research on the Social Context of Mathematics Education*, Windsor, NFER-Nelson.

COCKCROFT, W.H. (1982) *Mathematics Counts*, London, HMSO.

DESFORGES, C. and COCKBURN, A.D. (1987) *Understanding the Mathematics Teacher*, Lewes, Falmer Press.

DICKSON, L., BROWN, M. and GIBSON, O. (1984) *Children Learning Mathematics*, Eastbourne, Holt Education.

HART, K. (1981) *Children's Understanding of Mathematics: 11–16*, London, John Murray.

HOWSON, A.C. (1983) *A Review of Research in Mathematical Education, Part C, Curriculum Development and Research*, Windsor, NFER-Nelson.

SEWELL, B. (1981) *Use of Mathematics by Adults in Daily Life*, Leicester, ACACE.

SHUARD, H. (1986) *Primary Mathematics Today and Tomorrow*, London, Longman.

APPLYING THE RESEARCH MICROSCOPE

This section contains two papers, both of which involve precise examinations of aspects of the mathematics curriculum. Neville Bennett provides a masterly overview of research on the teaching and learning of mathematics in the primary school. His perspective — that of educational researcher — may not be very familiar to the teacher of mathematics. The chapter underlines the point made earlier that those of us in mathematics teaching and education have much to learn by looking outside our narrow domain. In the second paper Dietmar Kuchemann applies the results of the CSMS project (Hart, 1981) — to which he made some of the key contributions — in a critical examination of current school textbook treatments of ratio.

Reference

Hart, K. (1981) *Children's Understanding of Mathematics: 11–16*, London, John Murray.

9 *Teaching and Learning Mathematics in the Primary School*

Neville Bennett

In 1983 Charles Desforges and myself were asked by the Editorial Board of the *British Journal of Educational Psychology* to edit a monograph on *Recent Advances in Classroom Research* (Bennett and Desforges, 1985). In fulfilling this request we commissioned chapters in the major areas of classroom research from recognized leaders in the field. Reading these chapters created both optimism and concern: optimism, because of the clear elaboration of advances and refinement of theory and method; but concern, because of what appeared to be lack of communication between researchers operating within different content areas, or from differing methodological stances. It seemed that communication tended to be within special interest groups and/or through specialist conferences, rather than across the field as a whole. For example, mainstream researchers of classroom processes and mathematics educators were both carrying out studies utilizing similar foci and theoretical perspectives, but appeared not to know it.

The purpose of this chapter is two-fold: first, to overview very briefly the perspectives which have informed mainstream research on teaching-learning processes in natural classroom settings in order to identify the theoretical developments in the field over the last decade or so: second, to present findings from two of our recent studies which took a constructivist perspective on mathematics teaching and learning in primary classrooms.

Changing Perspectives

Research on teaching styles dominated the 1970s (Bennett, 1976; HMI, 1978; Galton *et al.*, 1980). This body of research indicated that more formal, didactic styles related to enhanced mathematics achievement (cf. Gray and Satterley, 1981), but the perspective contained several weaknesses. Not least was its inability to identify individual teacher activities or behaviours which related to higher maths achievement. As such it was of little value in initiating improvements in teaching. In addition, the differences in scores between styles were often relatively slight, which severely limited the use of style as an explanatory variable.

Dissatisfaction with the styles approach led to the search for alternative theoretical perspectives, a search which coalesced around the concept of opportunity to learn (Carroll, 1963; Bloom, 1976; Harnischfeger and Wiley, 1976). This perspective rejected the assumption underpinning the styles approach that a direct relationship exists between teacher behaviours and pupil learning. Instead it was argued that all effects of teaching on learning are mediated by pupil activities. In particular, the amount of time the pupil spends actively engaged on a particular topic is seen as the most important determinant of achievement on that topic. The measurement of this time is generally referred to as time on task, pupil involvement or engagement. In this approach the pupil is the central focus, with the teacher seen as the manager of the attention and time of pupils in relation to the educational ends of the classroom.

Research based on this approach has spawned an extensive literature since the mid-1970s, and studies still continue. With regard to mathematics the findings indicate that teachers spend on average some four and a half hours per week on the subject but that this varies from two to eight hours per week. In other words some pupils gain the opportunity to study maths three times more often than others. The time pupils spend actively involved in their work (i.e., on task) varies widely across classrooms and also across subjects, being lowest in maths and language activities (Bennett *et al.*, 1980).

The most consistently replicated findings link pupil achievement to the quantity and pacing of instruction. Specifically the amount learned is related to opportunity to learn, measured at its broadest by the length of school day, by the time allowed for the study of different subjects in the curriculum and by the amount of time pupils spend actively engaged on their tasks (Bennett, 1976, 1982; Brophy and Good, 1986).

In the United States, these findings have been developed into a prescriptive model of direct instruction from which teachers are urged to run structured, orderly, teacher directed classrooms with clear academic focus, frequent monitoring and supervision whilst maintaining a warm and encouraging climate. This model will not appeal to all teachers, but a very similar picture is portrayed by the most recent study of junior schools in Britain (ILEA, 1986).

The major limitation of the opportunity to learn approach is that time, or involvement, is a necessary but not sufficient condition for learning. Exhortations to increase curriculum time, or to improve levels of pupil involvement, are of no avail if the quality of the curriculum tasks themselves is poor, not worthwhile, or not related to children's attainments. Consequently contemporary thinking about teaching and learning has shifted the focus from time to the nature and quality of classroom tasks as they are worked under normal classroom conditions and constraints, i.e., to the interaction of teachers, pupils and tasks in complex social settings.

This shift in focus is reflected in recent professional concern about the appropriateness of tasks to children's attainments, which has centred on the concept of matching, i.e., the assignment of children to tasks which optimally sustain motivation, confidence and progress in learning. Teachers must '. . . avoid the twin pitfall of demanding too much and expecting too little' (Plowden, 1967). It is a recommendation easier to state than achieve as has been clearly demonstrated in a series of reports by Her Majesty's Inspectorate.

In a survey of over 500 primary schools in England, HMI concluded that teachers were underestimating the capabilities of their higher-attaining pupils, the top third of students in any class. In mathematics the the provision of tasks that were too easy was evident in one-half of the classroom observed. In later reports their concern has broadened to include low-attaining children. In two recent surveys of school catering for the age range 8 to 13 years they argued that both the more able and the less able were not given enough suitable activities in the majority of schools, and concluded that 'Overall, the content, level of demand and pace of work were most often directed toward children of average ability in the class. In many classes there was insufficient differentiation to cater for the full range of children's capabilities' (HMI, 1983, 1985). However, the data on which these findings are based are unsatisfactory from a research perspective. HMI's observation of the match of tasks and children were unstructured, unstated and undertaken without any clearly articulated view of learning or teaching.

Empirical research which has addressed this issue has been informed at a theoretical level by insights derived from cognitive psychology, and by theories of teaching which view classrooms as complex social settings. The adoption of cognitive psychological principles has moved the focus on learning from a behaviourist to a constructivist perspective. The assumptions underpinning this perspective are that the tasks on which pupils work structure what information is selected from the environment and how it is processed. Learners are not seen as passive recipients of sensory experience who can learn anything if provided with enough practice, rather they are seen as actively making use of cognitive strategies and previous knowledge to deal with cognitive limitations. In this conception learners are active, constructivist and interpretive, and learning is a covert, intellectual process providing the development and restructuring of existing conceptual schemes. As such teaching affects learning through pupil thought processes, i.e., teaching influences pupil thinking, pupil thinking mediates learning.

To understand learning thus requires an understanding of children's progressive performances on assigned tasks, and to understand the impact of teaching on learning it is necessary to ascertain the extent to which the intellectual demand in assigned work is appropriate or matched to children's attainments. Further, since classroom learning takes place within a complex social environment, it is necessary to understand the impact of social processes on children's task performances.

Doyle (1979, 1983) has produced the most elaborated model of classroom social processes. This views classrooms as complex social settings within which teachers and students are in a continuous process of adaptation to each other and the classroom environment. In Doyle's view the assessment system in operation in the classroom is at the heart of this process. Students must learn what the teacher will reward, and the teacher must learn what the students will deliver. Mutual accommodation leads to cooperation between teacher and taught, and in this theory cooperation, is the keystone to a classroom life acceptable to the participants. This perspective emphasizes the complex social interactions involved in classroom life, assigns a crucial role to the pupils in influencing the learning processes which teachers seek to manipulate, and places in central focus the role of assessment procedures. In this perspective studying matching as it actually occurs in classrooms entails observing which tasks teachers assign, how and why they assign them, how and why pupils interpret and work on them, and how and why teachers respond to pupils' work.

A Model of Task Processes

Research undertaken from this perspective in classrooms is of recent origin. Nevertheless, the findings produced have been very fruitful in guiding attention to significant features of teacher and pupil behaviour, and to aspects of classroom organization which have an impact on the quality of children's learning experiences in mathematics at primary level. In order to demonstrate this the findings of two of our recent studies are presented, one which investigated the mathematics teaching of 6 and 7-year-old children (Bennett *et al.*, 1984), and the other which contrasted learning experiences of junior age children in mixed and single age classes (Bennett *et al.*, 1987). These findings can be summarized around a model of classroom task processes presented in Figure 1.

Models are, by definition, simplified versions of reality, their role is to highlight the major influential factors in an area or process. Figure 1 thus highlights the major elements in classroom task processes as delineated by our research in this area. It conceives classroom task processes as cyclic. It assumes that teachers plan the tasks they will present to pupils, or will allow their pupils to choose, on the basis of clear and specific intentions, e.g. 'Angela needs work reinforcing symbol-sound relationships', or 'John is now sufficiently competent in the basic computation of Area and this should now be extended to applied or practical problems'. The tasks, once chosen, have to be presented to the child, group or class

Figure 1. A Model of Classroom Task Processes

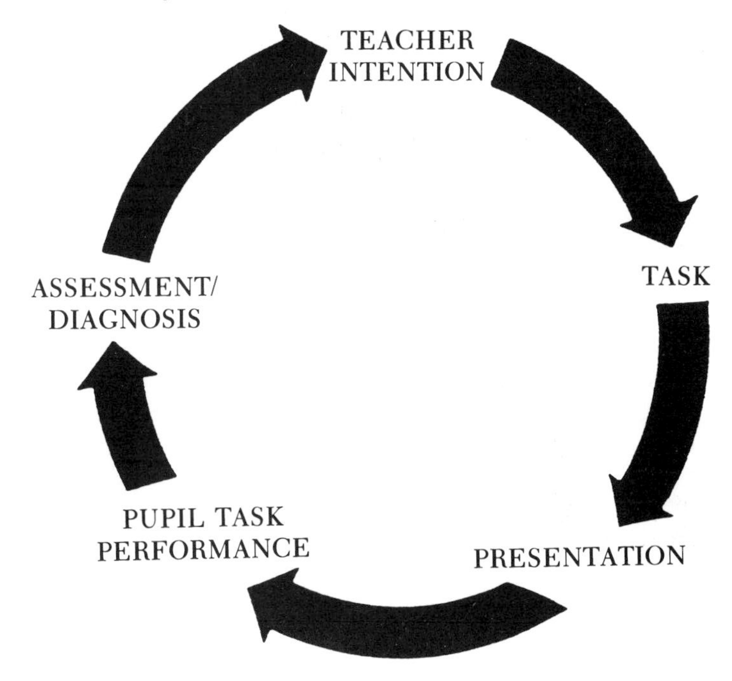

in some way. The presentation of tasks can take many forms, the major criterion being that children are clear what it is they are supposed to do. The pupils will then work on their tasks, demonstrating through their performances their conceptions and understanding of them. When the work is completed, it might be expected that the teacher will assess or diagnose the work in some way, and that the knowledge of the child's understandings gained would thereby inform the teacher's next intention.

This description is deceptively simple, however, since the possibility of a mismatch or an inappropriate link is apparent between every element of the model. These links are briefly considered below, drawing on the findings of our two recent studies.

Intention and Task

We define purpose or intention in terms of the intellectual demand that tasks make on learners. Drawing on Norman's (1978) theory of complex learning, five types of task demand were characterized as follows:

Incremental — introduces new ideas, procedures and skills
Restructuring — demands that a child invents or discovers an idea for him/herself

Enrichment — demands application of familiar skills to new problems
Practice — demands the tuning of new skills on familiar problems
Revision — demands the use of skills which have not been practised for some time.

Of the 212 maths tasks observed in our first study 44 per cent made practice, and 35 per cent incremental demands. There were virtually no restructuring or enrichment tasks (1 and 7 per cent respectively), a finding of some concern in the light of increasing demands for increased practical maths work. These proportions were almost identical for children irrespective of attainment level. As such low attainers received similar proportions of practice tasks as high attainers, a pattern likely to generate delays in progress for the latter and inadequate practice for the former. Further, the teachers' intended demands were not always those which children actually experienced, and this posed a particular problem for high attainers. Overall 30 per cent of all number tasks failed to make their intended demand, and high attainers suffered from this three times more often than low attainers.

Task and Pupil Performances

This is the link which HMI refer to as matching. The findings from both our studies support HMI judgments in this area. Table 1 shows the findings derived from the study of 6 and 7-year-olds, but the pattern was identical in our study of single and mixed age classes in junior schools. The table shows a clear trend for the underestimation of high attainers in the class and for the overestimation of low attainers. This trend, however, hides wide differences in the level of matching in different classrooms.

In tasks judged to underestimate children, performances were quick and accurate, although in many cases the teachers judged the work rushed and untidy. In most cases the production or recording routines limited the cognitive experience offered by the task. Very often, therefore, the children practised work with which they were already perfectly familiar, often a consequence of being directed to the next exercise in the scheme without adequate checking of their competence. The underestimation of high attaining children is serious, but perhaps less serious than the overestimation of low attainers since this could lead to blocks in learning or understanding with the possibility of decreased motivation. As such this is dealt with in more detail in what follows.

Among 6 and 7-year-olds output was low. On average children took three minutes to complete each calculation, on tasks which tended to focus on the four rules with quantities less than 20. Most effort was spent

Table 1. Matching of Number Tasks to Children of Differing Attainment Levels (percentages rounded)

Attainment	*Match*	*Too easy*	*Too hard*
High	41	41	16
Average	43	26	26
Low	44	12	44

Source: S.N. Bennett *et al. The Quality of Pupil Learning Experiences*, London, Erlbaum, 1984.

on the production features of tasks rather than on progress through an exercise, manifested in copying, rubbing out, boxing answers, and the like. The children's errors were often due to lack of understanding, or production errors. For example, one child was asked to complete simple division sums of the type $12 \div 3 =$ in order to consolidate this knowledge of the three times table. He only made one error but, when asked to recite the three times table in the post-task interview, he began '$3 \times 1 = 3$, $2 \times 5 = 10$, $4 \times 2 = 8$'. It transpired that he had completed the original task by copying all the answers from a wall chart illustrating multiplication and division by 3.

Examples of production errors occurred when children miscounted or misplaced cubes. One girl was asked to divide 48 by 3 using cubes. It took her a long time to collect sufficient cubes. She did not estimate what 48 cubes might look like but kept counting what she had and making persistent journeys for more. The cubes took up a lot of space, and several were knocked on the floor. She then allocated them one at a time to three piles. She did not know she was not starting with 48, and miscounting the quantities in the three piles she got three different answers. She was clearly puzzled and appeared to remain so.

From the tasks observed in the junior school samples four commonly occurring problems were:

1 difficulties in reading task instructions;
2 difficulties in understanding the required procedures;
3 use of inappropriate strategies;
4 insufficient knowledge to complete the task.

One child's task was to complete such mathematical sentences as:

7 tens and 2 =
2 tens and 11 =
2 tens and 11 = 3 tens and
3 tens and 17 =   tens and 7

A disc frame was available to help with this task. The child was unable to make any progress because of the difficulty he had in reading the question. He built up the words phonetically, but clearly had no comprehension of what was required. He attempted to use the frame for counting, but this activity was irrelevant to the question being attempted. He thus resorted to copying the questions into this book. No teacher contact was recorded throughout the period of this task.

Another child had a similar problem but in this instance received help from the teacher. He thus came to a partial understanding of his task which was to count straws into bundles of ten. Nevertheless, he spent much of the lesson gazing around the classroom, and at the end of the period had only managed to make one bundle of ten and one bundle of twenty straws.

Lack of ability to read or comprehend the tasks set was recorded in a number of tasks. One of the outcomes of this was the often enormous amount of time that children take to complete even simple tasks. One child who found difficulty in reading the task, which required him to arrange fifty-six straws into bundles of ten, took seventy-five minutes to read and complete this relatively simple exercise. Throughout this time he received no help or comment from the teacher.

On some tasks the demands of the procedure appeared to be a great deal more difficult than the demand of the actual question set. A typical example is of the child who was asked to undertake the conversion of '13 tens' into HTU. Using Dienes apparatus, he counted out with some difficulty, and many times, twenty-nine 'longs' and then laid these 'longs' on top of a 100 square ('a flat'). Having done this, he wrote '20'. He was then persuaded by the teacher to select a second 'flat', and then a third, laying the 'longs' on these. Eventually he wrote:

HTU
30

In addition to not understanding this procedure he encountered great difficulty in finding and retaining sufficient apparatus to carry out the procedure fully, in finding enough space on the table to lay it out, and in accurately accounting the 'longs' he had in front of him.

Sometimes the general procedure is understood at a concrete level which allows for no flexibility in its use. Thus one child knew the procedure for gaining a balance on a scale, but persisted slowly and conscientiously to use 2 gram weights irrepective of the size of the weight to be balanced.

This type of partial understanding of procedures is similar in kind to partial understanding of strategies. This was particularly noticeable in work on the four rules, where children appeared to have considerable problems in deciding which rule to apply. For example, one child was to undertake a task comprised of written problems which required one of the four rules to solve them. He experienced great difficulty in deciding which of the rules should apply. Instead of dividing he multiplied (using a number that did not appear in the question), and when faced with the question 'a runway is 3070 metres long, how many metres is this short of 5000 metres?', he attempted subtraction with the following configuration

$$\begin{array}{r} 3070 \\ -\ \underline{5000} \\ 2070 \end{array}$$

The worst cases of overestimation were where the child appeared to have no knowledge relevant to the task. One child's lack of knowledge of multiplication tables prevented her from doing anything other than copying out the first two questions. Even this was not begun until, after twenty minutes waiting, she checked with the teacher what she had to do. The task consisted of seven sums requiring the multiplication of a three-digit number by a single-digit number with a 'carry' figure. She was unable to even begin the task and as such was very little involved in it. For example, on six occasions she asked another child what time it was, and at the end of the lesson the work was put away uncompleted without being checked.

These case studies, presented to portray the reality of overestimation, do not reflect a peculiarly British phenomenon. The study by Anderson *et al.* (1984) in the United States provides very similar illustrations.

Teacher Diagnosis

Ausubel (1969) argues that the most important single factor influencing learning is what the learner already knows — 'ascertain this and teach him accordingly'. From our studies it is clear that teachers do not diagnose, i.e., attempt, by obervation or interview, to obtain a clear view of pupils' understandings and misconceptions. This is a serious omission since it simply stores up problems for later stages of children's learning.

The majority of the teachers observed, in both infant and junior classes, tended to be stationary at the front of the class, marking work while children queued for attention. Because of this pressure, the time spent with any individual child was short, and interactions not extensive. It could be argued that this reflects a justified pragmatic response strategy to the impossible situation of one adult being expected to provide high quality instruction appropriately matched to the individual capabilities of a large group of children. In order to cope in such a situation routines or procedures are brought into play which maximize efficiency at the expense of diagnosis and pupil understanding.

Lack of diagnosis was most often accompanied by teachers limiting their attention to the products of children's work rather than focusing on the processes or strategies employed by children in arriving at their product. When faced with pupil errors, teachers need to shift from a strategy which entails showing children how to do it to one which is exemplified by the request, 'show me how you did it'.

Lack of diagnosis means that teachers have insufficient knowledge of children's understandings to enable optimal decisions to be made concerning the intentions for the proceeding task. It is clear from our evidence that this in large part explains the provision of inappropriate tasks to children.

Conclusion

Current research on teaching-learning processes is focusing on the nature and quality of classroom tasks, the accuracy of diagnosis of children's understandings and misconceptions of concepts and content, and the quality of teacher explanations to this end. The centrality of these variables in effective teaching can be gauged from one of the conclusions of the House of Commons Select Committee Report which stated, 'the skills of diagnosing learning success and difficulty, and selecting and presenting new tasks are the essence of teachers' profession and vital to children's progress' (1986).

The approach takes due account of the role of the pupil in mediating and structuring knowledge, and places great stress on teachers' knowledge of subject matter, pedagogy and curriculum. These can be exemplified by questions of the following type: How can teachers teach well knowledge that they themselves do not thoroughly understand? How can teachers make clear decisions regarding what counts as development in curriculum areas with which they are not thoroughly conversant? How can teachers accurately diagnose the nature of children's misconceptions, and provide the necessary alternative learning experiences, without an adequate foundation of knowledge in the subject matter and associated pedagogy? These are important questions in relation to primary teaching, where teachers tend to be generalists and where worries are currently being expressed about the proportion of teachers who have difficulty selecting and utilizing subject matter, particularly in maths and science.

Another aspect of pedagogy which current research has identified as problematic, and which impinges on the quality of matching, is the organization and management of the classroom to provide optimal learning environments. The typical organization in maths is to mark work in front of children at the front of the class. This has unfortunate organizational consequences, including poor supervision of the rest of the class, queuing (sometimes at both sides of the desk), insufficient time adequately to diagnose children's learning and often teacher frustration. These factors, together with the role many teachers take on as the provider of instant solutions to a constant stream of problems, serve to create a learning environment which is far from optimal for teacher or taught.

Current work is addressing some of these issues. Studies are now underway on the characteristics of teacher knowledge which constitute cognitive skill in teaching in expert and novice teachers (Leinhardt and Smith, 1985) and on the manner in which teachers' knowledge of subject matter contributes to the planning and instructional activities of teaching (Shulman, 1986). Research on classroom management appears less strong but is crucially important. Attempts to confront the issue currently include the utilization of parental involvement in children's learning, and the use of cooperative grouping strategies. There is still much to be done, however, on short- and long-term effects and on implementation issues. Nevertheless, they hold out much promise for improving the quality of maths teaching and learning in the primary classroom.

References

ANDERSON, L.M., BRUBAKER, N., ALLEMAN-BROOKS, J. and DUFFY, G.S. (1984) *Student Responses to Classroom Instruction* (Final Report), Institute for Research on Teaching, Michigan State University.

AUSUBEL, D.P. (1969) *Educational Psychology: A Cognitive View*, New York, Holt, Rinehart and Winston.

BENNETT, S.N. (1976) *Teaching Styles and Pupil Progress*, London. Open Books.

BENNETT, S.N. (1978) 'Recent Research on Teaching: A Dream, a Belief and a Model', *British Journal of Educational Psychology*, 48, pp. 127–47.

BENNETT, S.N. (1982), 'Time to Teach: Teaching Learning Processes in Primary Schools', *Aspects of Education*, 27, pp. 52–70.

BENNETT, S.N. and DESFORGES, C.W. (1985) *Recent Advances in Classroom Research*, Edinburgh, Scottish Academic Press.

BENNETT, S.N., ANDRAE, J., HEGARTY, P. and WADE, B. (1980), *Open Plan Schools: Teaching, Curriculum and Design*, Slough, NFER.

BENNETT, S.N., DESFORGES, C.W., COCKBURN, S. and WILKINSON, B. (1984) *The Quality of Pupil Learning Experiences*, London, Erlbaum.

BENNETT, S.N., ROTH, E. and DUNNE, R. (1987), 'Task Processes in Mixed and Single Age Classes', *Education 3–13*, 15, pp. 43–50.

BLOOM, B.S. (1976) *Human Characteristics and School Learning*, New York, McGraw-Hill.

BROPHY, J.E. and GOOD, T.C. (1986), 'Teacher Behaviour and Student Achievement', in M.C. WITTROCK (Ed), *Handbook of Research On Teaching*, New York, Macmillan.

CARROLL, J.B. (1963) 'A Model of School Learning', *Teachers College Record*, 64, pp. 723–38.

DOYLE, W. (1979) 'Classroom Tasks and Students' Abilities', in P.L. PETERSON and H.J. WALBERT (Eds), *Research on Teaching: Concepts, Findings and Implications*, Berkeley, Calif., McCutcheon.

DOYLE, W. (1983) 'Academic Work', *Review of Educational Research*, 53, pp. 159–200.

GALTON, M., SIMON, B. and GROLL, P. (1980) *Inside the Primary School*, London, Routledge and Kegan Paul.

GRAY, J. and SATTERLEY, D. (1981) 'Formal or Informal? A Reassessment of the British Evidence', *British Journal of Educational Psychology*, 51, pp. 187–96.

HARNISCHFEGER, A. and WILEY, D.E. (1976) 'Teaching-Learning Processes in the Elementary School: A Synoptic View', *Studies of Education Processes*, No. 9, University of Chicago.

HER MAJESTY'S INSPECTORATE (1978) *Primary Education in England*, London, HMSO.

HER MAJESTY'S INSPECTORATE (1983) *9–13 Middle Schools: An Illustrative Survey*, London, HMSO.

HER MAJESTY'S INSPECTORATE (1985) *Education 8–12 in Combined and Middle Schools*, London, HMSO.

HOUSE OF COMMONS EDUCATION, SCIENCE AND ARTS COMMITTEE (1986) *Achievement in Primary Schools*, London, HMSO.

ILEA (1986) *The Junior School Project*, London, ILEA, Research and Statistics Branch.

LEINHARDT, G. and SMITH, D.A. (1985) 'Expertise in Mathematics Instruction: Subject Matter Knowledge', *Journal of Educational Psychology*, 77, pp. 247–71.

NORMAN, D.A. (1978) 'Notes towards a Complex Theory of Learning', in Lesgold *et al.* (Eds), *Cognitive Psychology and Instruction*, New York, Plenum.

PLOWDEN REPORT (1967) *Children and Their Primary Schools*, Report of the Central Advisory Council for Education (England), London, HMSO.

SHULMAN, L.S. (1986) 'Those Who understand: Knowledge Growth in Teaching', *Educational Researcher*, 15, pp. 4–22.

10 *Learning and Teaching Ratio: A Look at Some Current Textbooks*

Dietmar Küchemann

Students' Ratio Strategies

School students seem to show considerable resistance to adopting formal methods such as the rule-of-three for solving ratio and proportion tasks. Hart (1981), for example, who gave a written test on ratio to over 2000 English school students, reports that an inspection of the students' scripts revealed almost no evidence of the rule-of-three being used. Carraher (1986) interviewed seventeen Brazilian seventh grade students on a set of ratio tasks; only one student used the rule-of-three even though they had all received instruction on it during the seventh grade.

Instead of formal methods, school students seem to develop a host of informal methods, of varying degrees of effectiveness, for tackling ratio tasks. Figure 1 shows three tasks from the CSMS ratio test developed by Hart (1985). On task B, for example, Hart found that students may attempt to find the missing number of sprats by adding 1 to the original number of sprats; by adding 2 (which in this case gives a better answer); by doubling; or by adding 5 (because the length of the eel has increased by 5 (cm)). This last strategy has become known as the Addition Strategy (of which more shortly), though it is certainly not the only one that uses addition. Indeed, Booth (1981) makes the point that 'child methods' are typically additive, and in this particular case an additive approach can produce the required answer, by adopting an argument of this sort: to get from (10,12) to (15,?), add on half as much again, i.e. (10,12) + (5,6) = (15,18). Carraher (1986) calls this Rated Addition.

The Addition Strategy was identified by Piaget (see, for example, Inhelder and Piaget, 1958). It has since been extensively investigated by Karplus (for example, Karplus and Peterson, 1970), whilst Hart (1984) has devised teaching strategies that attempt to suppress it. Piaget regarded the strategy as an indicator of cognitive level, whilst Karplus has suggested

Figure 1. Three Ratio Tasks
Source: K. Hart, *Chelsea Diagnostic Mathematics Tests: Ratio*, Windsor NFER-Nelson, 1985.

Task A **Onion Soup Recipe for 8 persons**

8 onions
2 pints water
4 chicken soup cubes
2 dessertspoons butter
$\frac{1}{2}$ pint cream

I am cooking onion soup for 4 people.

How many chicken soup cubes do I need? ....

Task B There are 3 eels, **A, B** and **C** in the tank at the Zoo.

A 15cm long

B 10cm long

C 5cm long

The eels are fed sprats, the number depending on their length.

If **B** eats 12 sprats, how many sprats should **A** be fed to match?

A ....

Task C These 2 letters are the same shape, one is larger than the other. The curve AC is 8 units. RT is 12 units.

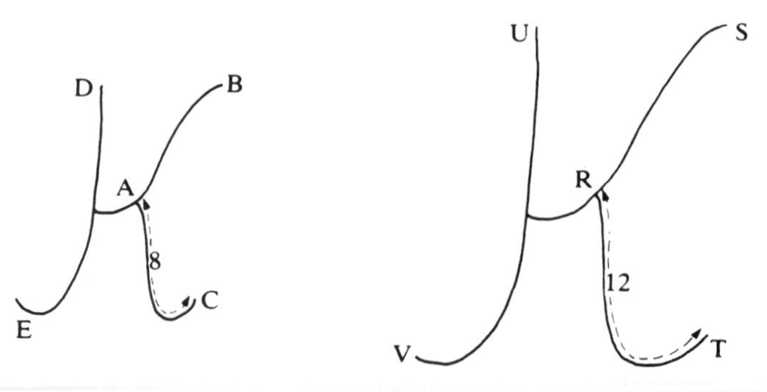

The curve AB is 9 units. How long is the curve RS? ....

Figure 2. Ratio Tasks against Facility
Source: K. Hart, *Chelsea Diagnostic Mathematics Tests: Ratio*, Windsor NFER-Nelson, 1985.

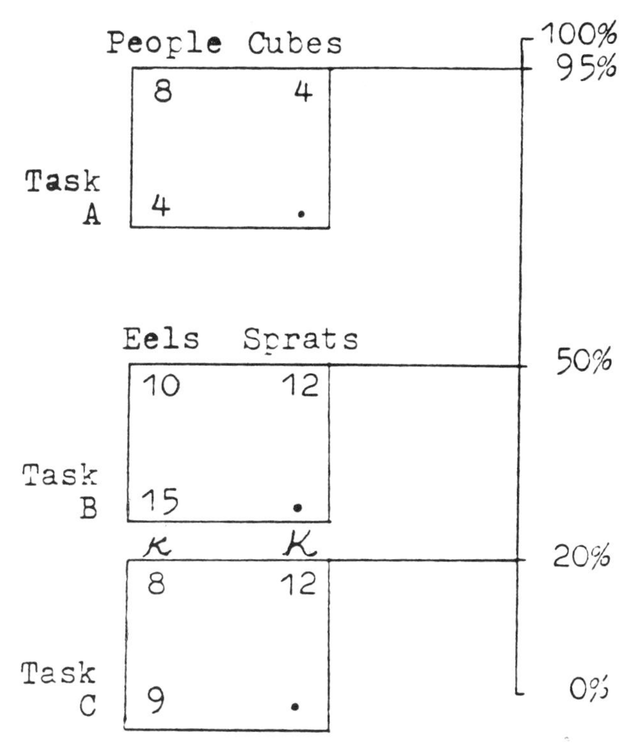

its use is more a matter of cognitive style (Karuplus *et al.*, 1974). Whichever view one may incline to, it is clear (see, for example, Küchemann, 1981; Karplus *et al.*, 1983) that the use of the strategy also depends on various characteristics of the given ratio task, in particular the numbers involved and the context.

In Figure 2 Hart's three ratio tasks are shown plotted against facility (obtained on a representative sample of 767 students aged about 14 years). In task C, 40 per cent of the sample gave the answer 13, commensurate with the Addition Strategy, whilst in task B only 9 per cent gave the corresponding answer of 17. Task A was answered correctly by 95 per cent of the sample, and it is doubtful that many of the remaining students used the Addition Strategy on this task as it gives the highly dissonant answer of 0 cubes.

The numbers in task A also lend themselves to the very simple, and in this case effective, strategy of halving. No such simple strategy works for B and C, though both can be solved by Rated Addition, in both cases by taking 'half as much again':

$$(10,12) + (5,6) = (15,18) \text{ and } (8,9) + (4,4.5) = (12,13.5).$$

Figure 3. Standard Photographic Print Sizes (Foto Inn)

COLOUR PRINTS (at time of processing)

Standard size prints	3½" × 5"	90 × 130mm	£0.18
*Post Card size	4" × 6"	100 × 150mm	£0.20
Enlargements	5" × 7"	130 × 180mm	£0.40
Enlargements	6" × 8"	150 × 200mm	£0.70
Enlargements	8" × 10"	200 × 255mm	£1.00

Why then did C prove so much more difficult than B, and why did so many more students resort to the Addition Strategy? The reason would seem to lie in the differences in context. In task C, increasing all the lengths in the small K by 4 cm can still produce a K that looks very much like the original: it could be difficult to *see* that the two Ks are not similar. On the other hand, the notion that a 5cm longer eel needs 5 extra sprats is in much more obvious conflict with the given information that a 10cm eel requires 12 sprats, not 10. This would perhaps be even clearer in the context of task A. This involves a soup recipe for 8 people, but consider the case for 9 people, say: the Addition Strategy argument, that the extra person requires an extra stock cube, strongly conflicts with the information that 8 people require only 4 cubes.

Task C involves geometic enlargement. Even in the case of rectilinear figures the Addition Strategy can seem plausible, as the standard sizes of photographic enlargements shown in Figure 3 testify.

Though some students may have a greater propensity towards using the Addition Strategy than others, the main point of the above has been to argue that the use of the Addition Strategy will also depend on the particular ratio task being tackled. The same argument can be made for the various other strategies in the individual student's repertoire (for example, Rated Addition is more attractive in task B than in C because it makes sense to say a 10cm eel plus a 5cm eel makes a 15cm eel, but not that two small Ks make a large one). As far as the teacher (and textbook writer) is concerned, this means students should be encouraged to engage in a wide variety of ratio tasks (in terms of number, context and overall difficulty) so that every opportunity is given for the various strategies that the student has constructed to emerge and to be made explicit — to the teacher as well as the student. In turn, this provides the opportunity for the strategies to be evaluated, be it to reconcile strategies, to (attempt to) eliminate them or to determine whether or when they are appropriate. Even if the teacher's aim is to help students adopt other, more powerful strategies (such as the rule-of-three), this is far more likely to be seen as worthwhile by the student if the limitations of his or her existing strategies are made explicit, rather than if the strategies are simply ignored (see for example, Case, 1978). As von Glasersfeld puts it, '. . . the child is unlikely to modify a conceptual structure unless there is an experience of failure or, at least, surprise at something not working out in the expected fashion' (von Glasersfeld, 1978, p. 14).

Crucially, the above approach is a constructivist one; it starts from where the students are, rather than with ideas and strategies that might hardly engage at all with the students' existing cognitive structures.

Ratio in Recent Mathematics Texts

Over the past few years I have been involved in a project that is producing mathematics textbooks for 11–16-year-olds. The first chapter devoted specifically to ratio (and proportion) comes in the year 2 book. In the upper track version (Harper *et al.*, 1987b) the chapter opens with a rather simple recipe task (in retrospect, perhaps too simple). The second page involves mixtures, and part of it is shown in Figure 4. The idea of using mixtures comes from a study by Noelting (1980), though he used mixtures of orange squash and water rather than tins of white and black paint. Mixtures do not, perhaps, embody the ratio relationship as strongly as do recipes of the kind in task A (the correspondence between people and stock cubes, say, seems very strong, whilst that between tins of white and black paint threatens to dissolve as the paints are mixed . . .). The attraction of mixtures, however, is the salience of the result, which can be expressed as a more primitive concept than it really is: 'Which tastes stronger?' 'Which is darker?' rather than 'Which has more orange squash compared to water?' 'Which has relatively more black paint?'

In Figure 4 students are deliberately not offered a method of solution. Rather they are expected to use methods of their own, which cannot be said of most of the extracts from other schemes considered later. From part b) onwards, students are given the opportunity to confront the Addition Strategy (though the strategy is not made explicit at this stage). Similar tasks, but in the context of coffee mixtures, are presented on the third page and also on the fourth, of which the lower track version (Harper *et al.*, 1987a) is shown in Figure 5. Again no specific method is put forward. Instead, by working With a Friend, it is hoped that students will begin to make their own methods explicit and to look at them critically. Part c) in particular, which is reminiscent of Bishop's celebrated fractions task (Lerman, 1983), provides considerable scope for students' strategies to emerge.

SMP 11–16 also uses the idea of paint mixtures. An extract from the booklet, *Ratio 1* (SMP, 1983a), which is intended for year 1 or 2 of secondary school, is shown in Figure 6. Mixtures of different strengths are presented, but students are not asked to compare them; this would be rather undemanding as the quantity of white paint is kept constant. Instead students are to determine different quantities of the mixtures. No

Figure 4. Extract from Harper et al. *NMP Mathematics for Secondary Schools 2 Red Track, Harlow, Longman, 1987*

B 1 a) Glenda and Basil are painting the garden shed.
Glenda's dad wants it grey.
The shop only has small tins of white paint and small tins of black paint.

Glenda mixes 2 tins of white paint with 7 tins of black.

Basil mixes 3 tins of white paint with 9 tins of black.

Will the two sides of the shed be the same colour?
If not, whose side will be darker?

b) Glenda adds one more tin of white and one more tin of black.
Does her colour get lighter or darker?

c) Glenda keeps adding one tin of each colour.
Does she ever get to Basil's colour?
If so, how many tins of each colour has she mixed?

2 to 7,
3 to 8,
4 to 9, . . .

specific method is spelt out, but the given numbers suggest that the authors had one method very much in mind (that of scaling-up by a whole number).

The first page of the SMP 11–16 ratio booklet is shown in Figure 7. Standard ratio notation is presented, which is then used to describe various (rather odd) situations. From the very start mathematical ideas are very clearly being imposed and, as with the other extract from the booklet, one feels that a number of opportunities have been lost: of gaining an insight into how students think, of connecting with their ways of thinking, and of offering a challenge. If von Glasersfeld is right, that children learn by encountering surprise or conflict, it is unlikely to happen here.

Some much more interesting tasks are to be found in the second SMP 11–16 ratio booklet (SMP, 1983b), under the heading 'Picture Puzzles', one of which is shown in Figure 8. It is a pity that something like this was not presented at the very beginning (though in a more appealing context and without the attempt to constrain students' thinking in part (a)). The tendency to *tell* students about ratio (rather than ask them), and to start with purely descriptive, and hence rather purposeless, tasks is not confined to SMP 11–16. Indeed, it seems almost universal. Thus it can be found in doughty revision series like *CSE Mathematics* (Greer, 1978) and *Basic Mathematics* (Elvin *et al.*, 1979), but also in more recent series like the *Integrated Mathematics Scheme* (IMS) (Kaner, 1982) and *Understanding Mathematics* (UM) (Cox and Bell, 1985ab).

Figure 5. Extract from Harper et al. *NMP Mathematics for Secondary Schools 2 Blue Track, Harlow, Longman, 1987*

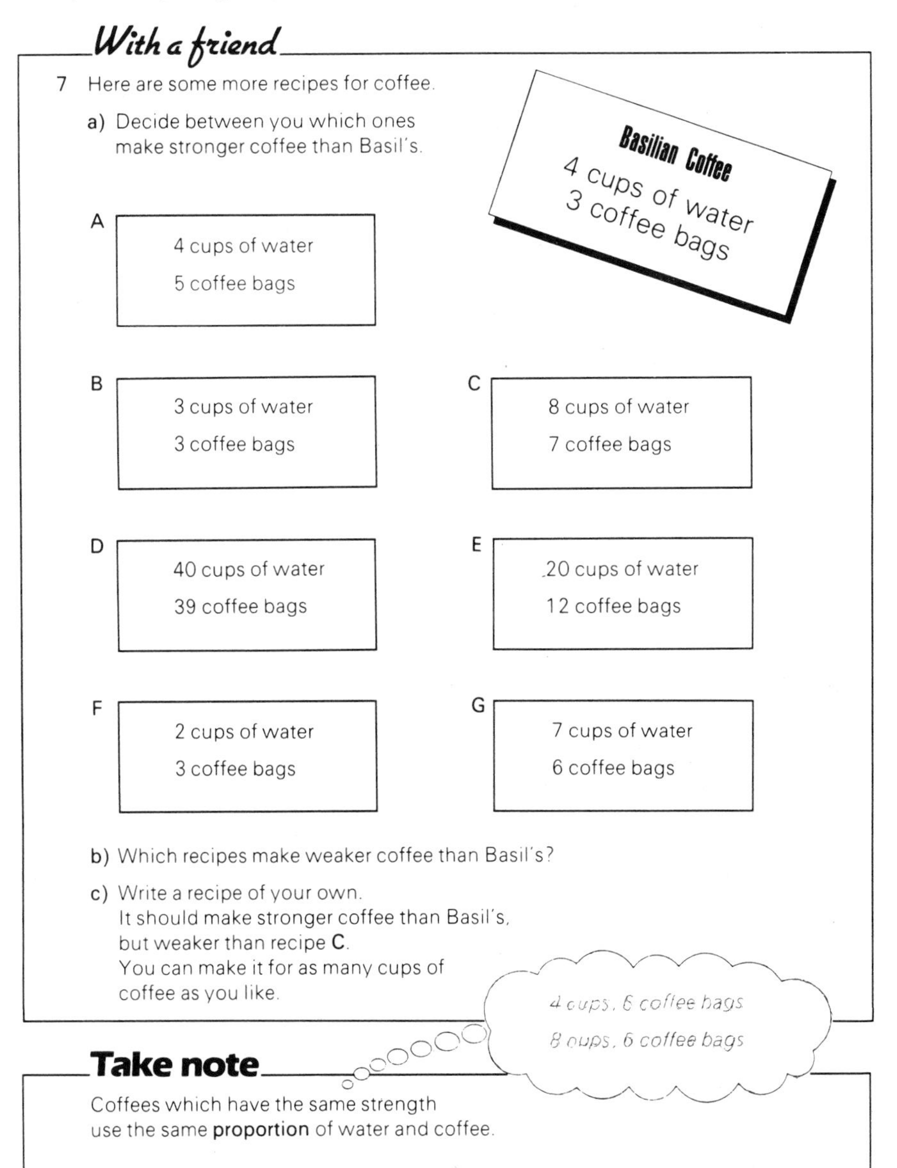

With a friend

7 Here are some more recipes for coffee.

a) Decide between you which ones make stronger coffee than Basil's.

Basilian Coffee
4 cups of water
3 coffee bags

A
4 cups of water
5 coffee bags

B
3 cups of water
3 coffee bags

C
8 cups of water
7 coffee bags

D
40 cups of water
39 coffee bags

E
20 cups of water
12 coffee bags

F
2 cups of water
3 coffee bags

G
7 cups of water
6 coffee bags

b) Which recipes make weaker coffee than Basil's?

c) Write a recipe of your own.
It should make stronger coffee than Basil's, but weaker than recipe **C**.
You can make it for as many cups of coffee as you like.

4 cups, 6 coffee bags
8 cups, 6 coffee bags

Take note

Coffees which have the same strength use the same **proportion** of water and coffee.

Figure 6. Extract from SMP, SMP 11–16, Ratio 1, Cambridge, Cambridge University Press, 1983.

F6 Here are the recipes for three different kinds of grey.

Extra light grey	**Very light grey**	**Light grey**
Mix black to white in the ratio 2 to 5.	Mix black to white in the ratio 3 to 5.	Mix black to white in the ratio 4 to 5.

(a) You are making extra light grey.
How many tins of white do you mix with 8 tins of black?

(b) You are making very light grey.
How many tins of black do you mix with 15 tins of white?

(c) You are making light grey.
How many tins of white do you mix with 16 tins of black?

(d) You mix 6 tins of black and 10 tins of white.
Which of the three kinds of grey do you get?

(e) You mix 12 tins of black and 30 tins of white.
Which of the three kinds of grey do you get?

Figures 9 and 10 show the beginning of the first ratio chapter in IMS and UM respectively. IMS actually starts with a definition of ratio which is both abstract and vague (compare it to the 'Take note' in Figure 5). Ratio notation is then introduced, simultaneously with the idea of simplifying ratios, and this is followed by a series of descriptive tasks. The UM approach is similar, though more gradual, with simplification of ratio deferred to the third page of the chapter.

In both chapters students are presented with a method for solving ratio tasks, and no thought seems to have been given to the ideas and strategies that the students may already have. In the case of IMS the method is none other than the rule-of-three (albeit made slightly simpler), which is introduced as early as the second page of the chapter (Figure 11). UM offers what is in effect the unitary method (Figure 12).

The unitary method is presented again at the start of the second year UM ratio chapter, together with a 'ratio method' (Figure 13). The latter is highly formal and sits oddly with the claim in the book's preface that 'the development of each topic was planned with reference to the findings of CSMS.' The chapter continues with an interesting discussion section, and it is a pity that this does not come first. In the exercise that follows many of the questions can be solved by Rated Addition, and it seems likely that many students will use this rather than the less accessible methods that they have been given.

Figure 7. Extract from SMP, SMP 11–16, Ratio 1, Cambridge, Cambridge University Press, 1983

A Ratio

In this picture,
each man has 2 horses.

We say

the ratio of horses to men is 2 to 1.

We also say

the ratio of men to horses is 1 to 2.

(a) The ratio of girls to dogs
is 1 to . . .

(b) The ratio of dogs to girls
is. . . to 1.

Figure 8. Extract from SMP, SMP 11–16, Ratio 2, Cambridge, Cambridge University Press, 1983

C2 Measure the height and width of this doorway.

(a) Copy and complete:

Height = ? × width

(b) **Which of the pictures A to G show the same doorway as this one?**

A

B

C

Figure 9. Extract from P. Kaner, Integrated Mathematics Scheme A2, London, Bell and Hyman, 1982

Unit M30 Ratio

Ratio of a pair of numbers

When two measurements in the same unit are to be compared we can use pure numbers to make the comparison. A pair of numbers used in this way is called a **ratio**.

Example 1:

In a club there are 20 girls and 40 boys.

The ratio of girls to boys is *20:40*.

This **simplifies** to 1:2 (dividing both parts of the ratio by 20).

Example 2:

The total surface area of the Earth is 197 million square miles.

139·4 million square miles are water and 57·5 million square miles are land.

The ratio of land to water is 57·5:139·4.

This simplifies to 1:2·4 (dividing both parts by 57·5).

Another way of saying this is, 'There is about two and a half times as much sea as land on Earth'.

Exercise M97

A What are the ratios of boys to girls in these situations?

1 In our class there are 17 boys and 19 girls.

2 In our club there are 25 boys and 40 girls.

3 In the swimming team there are 8 girls and 6 boys.

As with IMS, the first ratio chapter of the ST(P) Mathematics series (Bostock *et al.*, 1984) starts with notation and with the ideas of simplifying ratios. However, before embarking on the usual descriptive ratio tasks, two exercises on simplifying ratios and one on comparing ratios have to be gone through. This part of the chapter eschews virtually all context, and it is interesting to compare the worked example in Figure 14 with the coffee recipe tasks in Figure 5. Interestingly, the chapter includes an exercise (Figure 15) for which no specific method is advocated (the exercise is introduced by the statement, 'Some missing numbers are fairly obvious'). This allows students some scope for airing their own ideas, but it is not long before the rule-of-three and the unitary method are presented, both in a flurry of symbols (Figures 16 and 17).

In this brief look at published ratio materials I have focused on some of the more recent, more widely used secondary school mathematics schemes, and have looked only at the introductory work on ratio. It would

Figure 10. Extract from C.J. Cox and D. Bell, Understanding Mathematics 1, London, John Murray, 1985

12 Ratios; Proportional division

A Ratio, 1 : x

These gear wheels give a gear ratio of 1 : 2.

The large wheel has 24 teeth and the smaller wheel has 12 teeth, so the larger wheel turning once makes the smaller wheel turn twice.

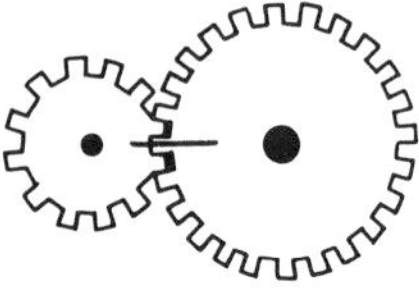

Fig. 12:1

Ratios can be written using the word 'to', using a colon (:), or as a fraction.

Example 12 to 24 = 1 to 2 = 1 : 2 = $\frac{1}{2}$

1 The ratio of John's money to Carol's money is 1 : 2. What is the ratio of Carol's money to John's?

2 Dad is 2 metres tall; Sam is 1 metre tall. What is the ratio of:
(a) Dad's height to Sam's height (b) Sam's height to Dad's height?

3 Write the ratio 1 to 3 as a fraction.

4

Fig. 12:2

CD is twice as long as AB, so CD : AB = 2 : 1 and AB : CD = 1 : 2.

EF is three times as long as AB, so EF : AB = 3 : 1 and AB : EF = 1 : 3.

Copy Figure 12:2, then write a sentence like the ones above about GH and AB.

be of interest to see how the ratio work is developed in the later books, though it seems unlikely that there would be any marked shift to a less didactic approach. I have argued that the materials need to engage with students' existing ideas and strategies, even if the aim is to move students to other ways of thinking. The materials that I have been involved in make a start in this direction, but the approach still seems to be rare, and in the schemes considered in this chapter almost non-existent.

Figure 11. Extract from P. Kaner, Integrated Mathematics Scheme A2, London, Bell and Hyman, 1982

Using a known ratio

If we know the ratio of two measurements and also one of the measurements we can find the other very easily.

Example 3:

The ratio of women to men in the army is 1:8. If there are 360 000 men, how many women are there?

1:8 must be the same as n:360 000 so n = 360 000 ÷ 8
= 45 000.

There are 45 000 women in the army.

Example 4:

In this picture the insect is enlarged in the ratio 3:10.

Measure the length of the insect in the picture and find the actual length of the insect.

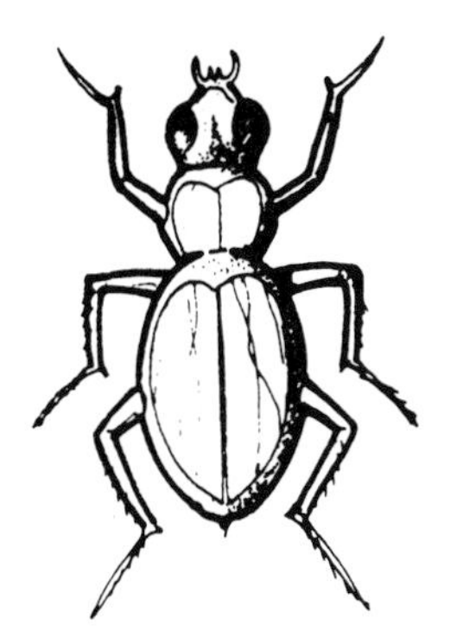

Length of picture 42 mm
Ratio 3:10 = 1:3·333
so 3:10 = length:42
1:3·333 = length:42
⇒ length = 42 ÷ 3·333
= 12·6 mm
Length of insect is 12·6 mm

Figure 12. Extract from C.J. Cox and D. Bell, Understanding Mathematics 1, London, John Murray, 1985

Using ratios

For Discussion

Fig. 12:11

Example Ann's pay to Tom's pay is in the ratio 4 : 5. Ann earns £100. What does Tom earn?

Divide Ann's pay into the 4 parts of her ratio, giving £25 a part. Tom receives 5 of these parts, giving: *Answer* 5 × £25 = £125.

Figure 13. Extract from C.J. Cox and D. Bell, Understanding Mathematics 2, London, John Murray, 1985

26 Ratios: proportion

Example 10 tins of food a week are required to feed 3 cats. How many tins would be required for 4 cats?

Ratio Method
The number of cats has increased in the ratio 4:3.
The number of tins must increase in the same ratio.
To increase in the ratio 4:3, multiply by $\frac{4}{3}$.
Answer: $10 \times \frac{4}{3} = 13\frac{1}{3}$ tins.

Unitary ('one') Method
3 cats require 10 tins
1 cat requires $\frac{10}{3}$ tins
4 cats will require $4 \times \frac{10}{3} = 13\frac{1}{3}$ tins.

Questions for Discussion

(a) Three eggs cost 15p. What do eight eggs cost?

(b) Three packets weigh 5 kg. What will seven of the same packets weigh?

(c) A spring increases by 36 mm with a 7.2 kg load. What is the increase with a 9.2 kg load?

(d) Four men on a raft have enough food for 21 days. If three more survivors are picked up, how long will the food last at the same rate?

(e) £14 will feed two cats for 12 weeks. For how long will £21 feed them? If a stray joins the two cats, for how long will the £14 feed them?

1 A bus travels 100 km in 4 hours. How far would it travel in 3 hours?

2 A man pays the same tax each week. He pays £500 in 8 weeks. How much will he pay in 10 weeks?

3 Six exercise books cost £1. What will fifteen exercise books cost?

Figure 14. Extract from L. Bostock et al., *ST(P) Mathematics 2, Cheltenham, Stanley Thornes, 1984*

Which ratio is the larger, 6 : 5 or 7 : 6?

(We need to compare the sizes of $\frac{6}{5}$ and $\frac{7}{6}$ so we express both with the same denominator.)

$$\frac{6}{5} = \frac{36}{30} \quad \text{and} \quad \frac{7}{6} = \frac{35}{30}$$

so 6 : 5 is larger than 7 : 6

Figure 15. Extract from L. Bostock et al., *ST(P) Mathematics 2, Cheltenham, Stanley Thornes, 1984*

Find the missing numbers in the following ratios:

1. 2 : 5 = 4 :

2. : 6 = 12 : 18

3. 24 : 14 = 12 :

4. $\frac{6}{\ } = \frac{9}{3}$

5. 3 : = 12 : 32

6. : 15 = 8 : 10

7. 9 : 6 = : 4

8. $\frac{\ }{4} = \frac{15}{10}$

9. $\frac{6}{8} = \frac{\ }{12}$

10. 6 : 9 = 8 :

Figure 16. Extract from L. Bostock et al., *ST(P) Mathematics 2, Cheltenham, Stanley Thornes, 1984*

A book of 250 pages is 1.5 cm thick (not counting the covers).

a) How thick is a book of 400 pages?

Method 1 (using algebra):

a) If the second book is x cm thick, then $\frac{x}{1.5} = \frac{400}{250}$

$$1.5 \times \frac{x}{1.5} = 1.5 \times \frac{400}{250}$$

$$x = 2.4$$

The second book is 2.4 cm thick.

Figure 17. Extract from L. Bostock et al., *ST(P) Mathematics 2, Cheltenham, Stanley Thornes, 1984*

Method 2 (unitary method):

a) 250 pages are 15 mm thick

1 page is $\frac{15}{250}$ mm thick

so 400 pages are $\frac{15}{250} \times 400$ mm thick

that is, 24 mm or 2.4 cm thick

References

BOOTH, L. (1981) 'Child Methods in Secondary Mathematics', *Educational Studies in Mathematics*, 12, pp. 29–40.

BOSTOCK, L., CHANDLER, S., SHEPHERD, A. and SMITH, E. (1984) *ST(P) Mathematics 2*, Cheltenham, Stanley Thornes.

CARRAHER, T. (1986) 'Rated Addition: A Correct Additive Solution for Proportion Problems', *Proceedings of the Tenth International Conference of Psychology of Mathematics Education*, London, University of London Institute of Education.

CASE, R. (1978) 'A Developmentally Based Theory and Technology of Instruction', *Review of Educational Research*, 48.

COX, C.J. and BELL, D. (1985a) *Understanding Mathematics 1*, London, John Murray.

COX, C.J. and BELL, D. (1985b) *Understanding Mathematics 2*, London, John Murray.

ELVIN, R., LEDSHAM, A. and OLIVER, C. (1979) *Basic Mathematics: Revision and Practice*, Oxford, Oxford University Press.

GREER, A. (1978) *CSE Mathematics 1*, Cheltenham, Stanley Thornes.

HARPER, E. *et al.* (1987a), *NMP Mathematics for Secondary Schools 2 Blue Track*, Harlow, Longman.

HARPER E. *et al.* (1987b), *NMP Mathematics for Secondary Schools 2 Red Track*, Harlow, Longman.

HART, K. (1981) 'Ratio', in K. HART (Ed.), *Children's Understanding of Mathematics 11–16*, London, John Murray.

HART, K. (1984) *Ratio: Children's Strategies and Errors*, Windsor, NFER-Nelson.

HART, K. (1985) *Chelsea Diagnostic Mathematics Tests: Ratio*, Windsor, NFER-Nelson.

INHELDER, B. and PIAGET, J. (1958) *The Growth of Logical Thinking from Childhood to Adolescence*, London, Routledge and Kegan Paul.

KANER, P. (1982) *Integrated Mathematics Scheme A2*, London, Bell and Hyman.

KARPLUS, R. and PETERSON, R.W. (1970) 'Intellectual Development Beyond Elementary School II: Ratio, a Survey', *School Science and Mathematics*, 70, 9, pp. 813–20.

KARPLUS, E.F., KARPLUS, R. and WOLLMAN, W. (1974) 'Intellectual Development Beyond Elementary School IV: Ratio, The Influence of Cognitive Style', *School Science and Mathematics*, 74, (October 1974), pp. 476–82.

KARPLUS, R., PULOS, S. and STAGE, E.K. (1983) 'Proportional Reasoning of Adolescents', in R. LESH and M. LANDAU (Eds), *Acquisition of Mathematics Concepts and Processes*, London, Academic Press.

Küchemann, D.E. (1981) 'Cognitive Demand of Secondary School Mathematics Items', *Educational Studies in Mathematics*, 12, pp. 301–16.

Lerman, S. (1983) 'Problem-solving or Knowledge-centred: The Influence of Philosophy on Mathematics Teaching', *Int. J. Math. Educ. Sci. Technol.*, 14, pp. 59–66.

Noelting, G. (1980) 'The Development of Proportional Reasoning and the Ratio Concept I — Differentiation of Stages', *Educational Studies in Mathematics*, 11, pp. 217–53.

SMP (1983a) *SMP 11–16, Ratio 1*, Cambridge, Cambridge University Press.

SMP (1983b) *SMP 11–16, Ratio 2*, Cambridge, Cambridge University Press.

Von Glasersfeld, E. (1987) 'Learning as a Constructive Activity', in C. Janvier (Ed.), *Problems of Representation in the Teaching and Learning of Mathematics*, Hillsdale, N.J., Erlbaum.

QUESTIONING THE SACRED COWS

Enthusiasm — passion even — is a good thing in education. But one of the dangers in mathematics teaching and education is to become infected with an uncritical, missionary zeal. The 1960s saw it: apparatus, activity and discovery learning in mathematics put on a pedestal. The 1980s have also seen it: practical work, problem-solving, investigations, cooperative groupwork and discussion all elevated as the ultimate 'goods' (or even 'gods') of mathematics teaching. The current orthodoxy takes paragraph 243 of the Cockcroft Report (1982) as the text for its sermon, full of exhortations to 'improve'. Practical work, problem-solving, investigations and discussion have become the sacred cows of mathematics teaching, and to question their worth is to invite the wrath of the believers. It is not for nothing that the ESG advisory teachers have been termed the 'Cockcroft Missionaries'!

It is not the aim of this section to attack these sacred cows or shibboleths of mathematics teaching, but rather to urge caution. If we give up our critical faculties, we give up a crucial means of improving practice. Investigative mathematics, as I have argued elsewhere (Ernest, 1984), offers a key means of developing student mathematical processes and confidence. But poorly designed investigative lessons can be as bad as the worst rote learning lessons. I have seen children 'investigating' meaningless numerical situations with no understanding of the task or its significance, and leaving the lesson after having been told the 'right' answers, with no sense of achievement or enjoyment, let alone any worthwhile intellectual synthesis. In Chapter 6 Steve Lerman warns of how the investigations so enthusiastically endorsed by zealots become assimilated into teachers' previous approaches. Investigations too can have right or wrong approaches and right or wrong answers, if taught in the traditional way. Even when used creatively, investigations can be bolted-on extras on the margins of the mathematics curriculum. Finally, a fully fledged investigational approach to mathematical topics in the mainstream of the curriculum can also be employed ineffectually. Investigations are no more a guarantee of success in mathematics teaching than any other approach, as Charles Desforges argues rather forcefully in his book (Desforges and Cockburn, 1987).

This section looks critically at two of the sacred cows of mathematics teaching. Charles Desforges considers discussion in primary mathematics, and finds that it does not always live up to the expectations of enthusiasts. Kath Hart examines the impact of practical experience on the formal understanding of the same mathematical topic, and finds that the orthodox view is not borne out in practice. Children seem to make less use of the concrete experiences we arrange for them, as a basis for their formal understanding, than our expectations and rhetoric would allow.

The lesson to be learnt from this is not that practical work, problem-solving, investigations and discussion should be abandoned. Rather that these very worthwhile approaches must be used carefully and critically. With these less directive teaching methods we must be very clear about our aims, and make sure that the experiences really are contributing to the children's growth in understanding.

References

Cockcroft, W.H. (1982) *Mathematics Counts*, London, HMSO.

Desforges, C. and Cockburn, A.D. (1987) *Understanding the Mathematics Teacher*, Lewes, Falmer Press.

Ernest, P. (1984) 'Investigations', *Teaching Mathematics and Its Applications*, 3, 3, pp. 80–6.

11 *There Is Little Connection*

Kath Hart

The research project 'Children's Mathematical Frameworks' (CMF) was financed by the ESRC at Chelsea College during the years 1983–85. It followed and built upon the research projects 'Concepts in Secondary Mathematics and Science' (CSMS) and 'Strategies and Errors in Secondary Mathematics' (SESM). In the first a description of levels of understanding in ten topics commonly taught in the secondary school was formulated from the data obtained from both interviews and more formal testing. The test results showed that many pupils committed the same error when attempting certain questions. The reasons for some of these specific errors were investigated in the subsequent research (SESM). These misconceptions do not begin at the secondary school stage but are formed earlier. A possible learning experience during which such misconceptions can arise is when the child is required to move from a practical or material-based approach to mathematics to the formal and symbolic mathematical language used in the secondary school. CMF was designed to monitor this transition to formalization in a number of topics taught to children aged 8–13 years.

The investigation involved many one-to-one interviews with children as well as observation of them being taught mathematics in their normal classrooms. Volunteer teachers (often those attending Diploma or Masters degree courses in mathematics education) were recruited and asked to select a topic from the mathematics curriculum scheduled for their class and involving the transition from practical work to formal mathematics. The formalization was a rule or formula verbalized and often written symbolically as a result of the structured work which had gone before. An example of such a formalization can be found in the teaching of area. Many primary school teachers lead up to the formula for the area of a rectangle by providing work which requires the child to (1) fill space with squares, (2) record the number of units in the length and width of a rectangle, (3) draw up a table which shows length, width and area, and (4) from it deduce a relationship which can be formalized as $A = l \times w$. The formula is an example of 'formalization', and the preceding experiences with squares and rectangles form part of the 'practical work'. The teacher, having chosen a topic, was asked to write a scheme of work which detailed the approach to the teaching, the equipment, work cards and the situations the children would meet. The interview cohort was chosen by the teacher from the class being taught. Sometimes, when the teacher was dealing with a class of thirty to forty pupils, the cohort was chosen to represent different levels of attainment; on the other hand, if a group of children had been withdrawn because they were assumed to be 'ready' for the formalization, then six of these were interviewed. Each child was interviewed four times, before the teaching started, just before and immediately after the lesson in which the formalization was verbalized, and then three months later. The questions which the children were asked were designed to test (1) their closeness to verbalizing the rule, (2) the acceptance of the rule as the 'preferred' method, and (3) the efficient use of the rule or formula. Very often the teacher stressed other (and more fundamental) mathematical concepts in the process of teaching the rule, so questions were asked about these.

The results will be published in book form, and a suitable subtitle might be 'Sums Are Sums and Bricks Are Bricks', which accurately describes the lack of connection between the two types of learning made by the children. When asked for the connection between practical work and symbolic statement of rule, the children's best reply was that one was a quicker route to the answer than the other. Nobody mentioned that the practical experience provided the data on which the formula was built. The teachers did not stress why this procedure was being followed, nor emphasize the generalizability of the rule and thus the advantage of accepting it. Often part of the philosophy underlining the teaching methodology which suggests there be practical experiences leading to a formalization is that the children themselves will be ready to notice and verbalize the relationship apparent from their work. The teacher draws the attention of others to the phenomenon, and perhaps couches the relationship in elegant terms. This presupposes some uniformity of progress amongst the pupils. One of the topics investigated was the substraction algorithm with decomposition, and two groups of eight children were withdrawn from their classes and taught by the same teacher. The teacher chose these pupils because they were considered to be at about the same level of attainment and 'ready' for the work. Figure 1 shows for these two groups the results obtained from the four interviews. One axis shows methods used by the children when presented with a subtraction question such as

$$\begin{array}{r} 246 \\ -\ \underline{158} \\ \underline{} \end{array}$$

Note that some children were already using the common incorrect method of always subtracting the smaller digit, whether it was on the bottom or

Figure 1. Children's understanding of Subtraction during Four Interviews

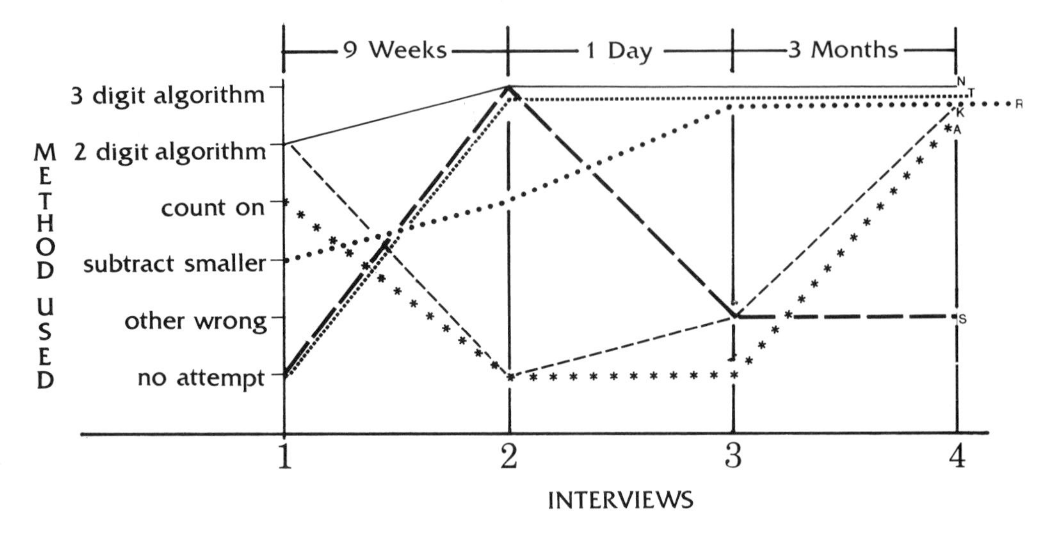

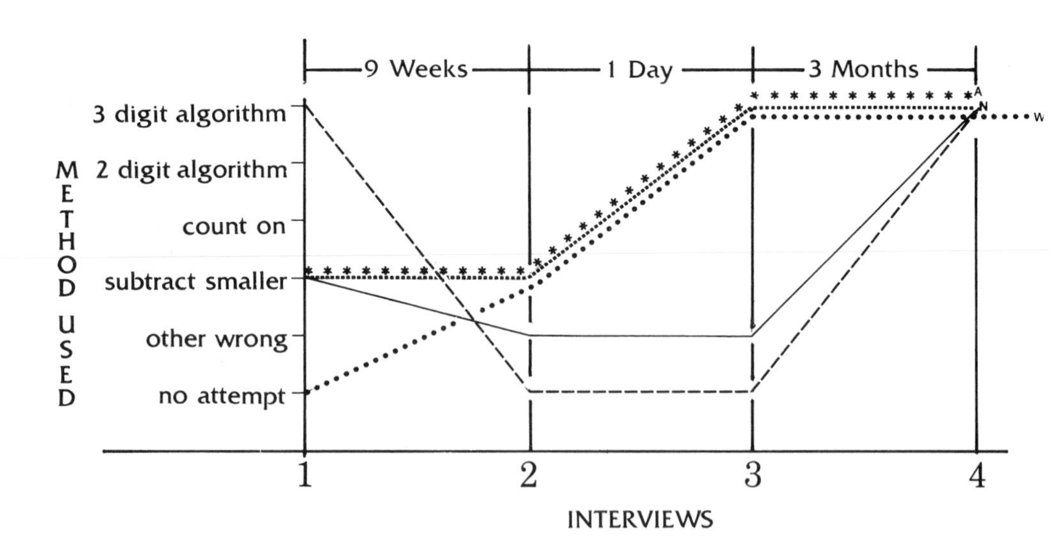

top line, before the teaching started. Additionally there were children who had a workable method of 'counting on' prior to the teaching but who could not attempt the questions after the formalization.

The diagrams show clearly that children chosen as being at the same stage of readiness, and taught by the same teacher for the same length of time, displayed very different patterns of progress. Although they nearly all appear to have achieved an ability to use the algorithm after three months, their responses to a question on decomposition display faulty reasoning. At the interview three months after the teaching the children were asked to do a subtraction question, and then by looking at the 'working out' to say what number was on the top line. Further questioning probed whether they thought the top line number had changed in value; five of the six children asked this question thought it had changed in value. During the observed lessons some teachers appealed to the evidence provided by the configuration of bricks or discs as superior or more convincing than the formula just taught. 'Now, we've actually done it, in front of ourselves, with the material, we've seen it work — you've done it. There shouldn't be any mystery about it because you're the ones who've actually moved the bits and pieces around.' Additionally children were encouraged to return to the materials if they were having difficulty solving problems using the taught rule. This implies that the child can provide the model to match the rule. If we consider the case of the teaching of equivalent fractions, one of the research classes used both discs and a fraction wall to build up a symbolized set of pairs of equal fractions. The teacher told the children what to use as a whole (usually a representation of 12) and then what parts to find. None of the three teachers observed teaching equivalent fractions provided a general method for setting up materials to find *any* pair of equivalent fractions. Indeed, to do so is rather a difficult task and it may be that one can only set up a convincing demonstration using bricks if one already knows the answer. This seemed true of the demonstrations of equivalence using sections of a circle. Both teachers and children drew diagrams such as those shown in Figure 2 which are inadequate for *demonstrating* the equivalence but give a spurious veracity if one already knows the fact.

Referring children to materials because they are having trouble with a formula is of little help unless one is sure that they have a workable method of setting up objects in order to mirror symbols. The theories of Piaget have influenced the teaching of mathematics to young children, and these theories have been translated in many cases to a belief that 'practical work' is a 'good thing'. The results of the CMF investigation show that we need to think carefully about the assumptions we make concerning the transition from practical work to formalization, and whether the methods employed when using material really are translatable into the terms of the algorithm. Consider, for example, the case of a class of 8-year-olds who used Unifix blocks to do subtraction questions building up to the algor-

Figure 2. Mark's Cakes (to check 3/8, 10/26)

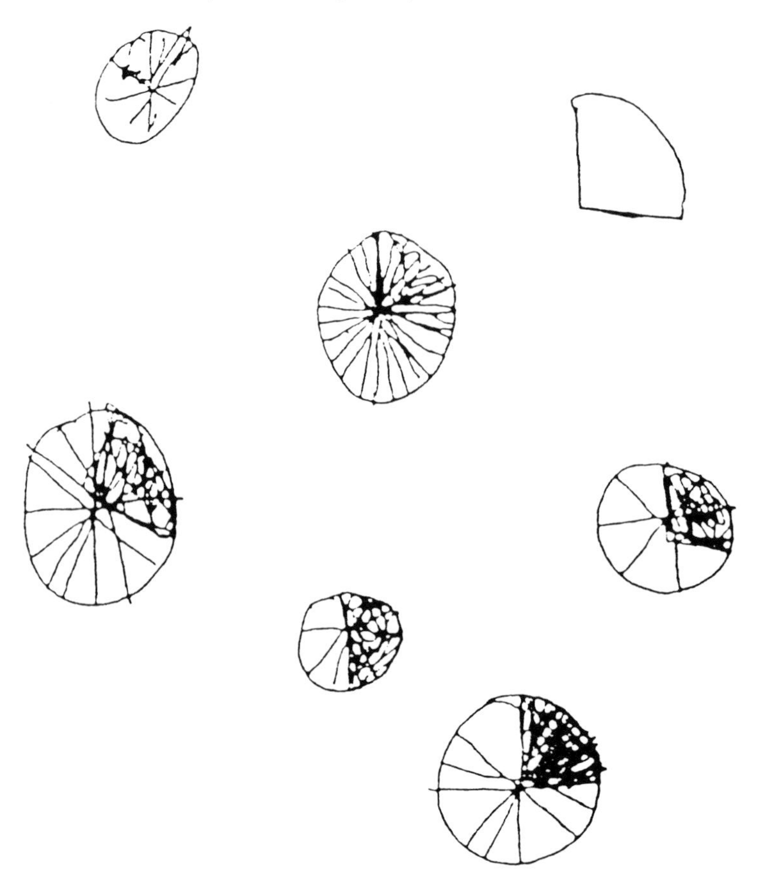

ithm involving decomposition. A valid and much used method of solving 56 — 28 was to set out 56 as five columns of ten bricks and six single bricks and then to use 28 as a mental instruction. This was followed by the removal of three of the tens, returning two units (broken off one of the tens) to the table. Finally, the collection still left on the table was counted. This is an adequate way of dealing with subtraction, but it has very little connection to the algorithm which is supposed to result from all the experience with bricks.

Many of us have believed that in order to teach formal mathematics one should build up to the formalization by using materials, and that the child will then better understand the process. I now believe that the gap between the two types of experience is too large, and that we should investigate ways of bridging that gap by providing a third transitional form. Nuffield Secondary Mathematics has a one-year research grant to investigate possible transitional experiences, and successful outcomes will become part of the information given to teachers in the Teachers' Resource Material.

12 *Classroom Processes and Mathematical Discussions: A Cautionary Note*

Charles Desforges

Classroom discussion features prominently in the modern rhetoric of mathematics education. As part of a programme to take children beyond competence in basic skills and towards confidence and success in applications work, discussion, both between teacher and children and amongst children independently of their teachers, is considered to play a valuable part. Through discussion it seems children should learn to articulate their points of view, to listen to others, to learn to ask appropriate questions, to learn how to recognize and respond to mathematically relevant challenges and in these ways to develop their mathematical conceptions and their applications.

Surprisingly, given the prominence, the empirical basis and the theoretical rationale on which this view is based are rarely spelled out. The value of discussion, it seems, is taken to be self-evident. This is possibly part of the modern vogue in which children are seen as 'social constructionists' building their understanding of the world by reflecting on their social experiences which take a predominantly linguistic form. Or perhaps it is just plain common sense. Whatever its provenance, the merits of discussion in mathematics are widely extolled. Cockcroft (1982) suggested conversations to be a key element in the mathematics curriculum. Some enthusiasts go further and consider that such exchanges should be at the heart of children's classroom mathematics experience (Easley and Easley, 1983).

The Easley's rest their enthusiasm in part on a constructivist philosophy of cognitive development and in part on their interpretation of four months' observations of the technique in practice in a Japanese elementary school. The Easleys report that the teachers at the Kitameo School seemed to place their highest priority on teaching children how to study mathematics largely through group work. Their normal technique was to ask the children to work individually on a challenging problem and then have them discuss their answers and reasons with other members of

Figure 1. Elementary Addition Task

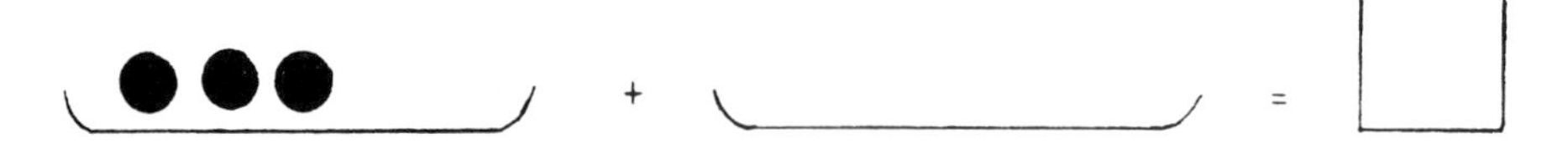

a mixed ability group to try to achieve a consensus of opinion. The quality of these exchanges is illustrated by the Easleys in an example in which the children were discussing the illustration in Figure 1.

Child 1:	You can't add three and zero. There is no answer.
Teacher:	Do you mean the answer is zero?
Child 1:	Yes.
Teacher:	Who agrees?
Child 2:	You can't add three and zero so your answer is three.
Teacher:	You can't add or there is nothing to add? Which is it?
Child 2:	There is nothing to add.
Child 3:	There are three here, even if there are none here, there are three.

What is impressive about this exchange is that it is unimpressive. Given the Easleys' enthusiasm and their desire to proselytize, this excerpt from their observations might be taken to be exemplary. Yet it contains no pupil-pupil exchanges. Nor is it at all clear what the children are making of the various contributions. The Easleys' example makes poor advertizing copy.

With material like this it is perhaps no surprise that the use of discussions in mathematics has yet to make much of an appearance in primary schools. HMI persistently complain about a preponderance of pencil and paper work in maths. Their observations have been supported by independent studies. Bennett *et al*. (1984) recorded no instances of discussion in more than 300 mathematics tasks observed in over two terms in thirty-two primary classes. Recent research by me and Anne Cockburn confirms that mathematical discussions are at a premium (Desforges and Cockburn, 1987). The same research, however, suggests that this has less to do with the advertizing than with the product: discussion causes a large number of problems for teachers and is apparently not well adapted to classroom conditions. In the rest of this chapter I illustrate and expand on these problems and consider their implications for the development of primary mathematics teaching.

The Research

A great deal of work has been done on understanding children's mathematical thinking, on identifying and sequencing educational objectives and on designing attractive teaching materials. The work has generated plenty of advice to teachers. Most of this advice does not seem to be practised in most classrooms.

The research reported here was part of an attempt to understand teachers' classroom behaviour (Desforges and Cockburn, 1987). It was based on the view that since teachers in our experience seem to know, understand and accept the precepts of good mathematics teaching and yet do not practise them, and that since most teachers are industrious and have the best interests of children at heart, there were probably some good reasons why traditional practices are resistant to exhortation. We argued that since familiar forms of curriculum development have had little impact on mathematics teaching it was necessary, before further attempts at change were made, to understand the forces that constrain teachers' actions.

To this end we recruited a sample of seven primary school teachers who had good reputations. Preliminary interviews established that the teachers held elaborate views of children's learning and recognized the virtues of the broad range of teaching methods described in the Cockcroft Report. They subscribed to the view that teaching methods should make the most of young children's well developed thinking capacities.

The teachers were each observed teaching mathematics over a period of three weeks. A sample of their teaching was video-recorded. The recordings were used to stimulate the teachers' recall of their thinking during teaching. The teachers were invited to stop the video whenever they felt a significant event had occurred or a decision had been made. The discussions about the tapes were held on the same day as the lesson was recorded. They were tape-recorded and subsequently analyzed to identify the teachers' experience of classroom interactions. The following comments apply only to the teachers' accounts of their thinking during discussion. These sessions accounted for less than 10 per cent of the sessions videoed.

Classroom Processes

The teachers reported a large number of tensions experienced whilst conducting discussions. Some of these are illustrated in the following example. Here the teacher set out to get a group of 6-year-olds to discuss different ways of making ten. The extract below is a record of one minute of her session.

Mrs D: Make a stick of ten cubes that are all the same colour. [Children do so] Right. Check you've got ten. Now. Show me with your cubes a number story that adds up to ten.

[The children break up their sticks and mutters of 'one add nine', 'two add eight' can be heard. Ross breaks his into two fives, looks round, rebuilds his ten and breaks it into one and nine.]

Mrs D: No Ross. You were right. I just said, 'show me a number story'.

[Ross reproduces two sticks of five.]

Mrs D: Now. What have you got, Ross?

Ross: Five add five.

Mrs D: What have you got Hazel?

Hazel: One add nine.

Mrs D: What have you got Kai?

Kai: Three add six equals ten.
[Kai is holding up a stick of six and a stick of four.]

Mrs D: Nooo . . . have another look.

Kai: Four add six.

Robert: Two add seven.

Mrs D: [Waving away Freddie from another group.] You count again. What do you think they are Darren?

Darren: Three add eight.

Before examining the teacher's experience of this brief exchange it is worth noting that she had thirty-five children in a very small room. She was working with one of three groups. Her comments on the action were as follows:

> I saw Ross do something unusual in making five add five. As soon as someone said 'one add nine' he [Ross] started to put his back. I was also trying to see what everyone else had done and who was looking at whose. Louise, Hazel and Lindsey are never too sure. I saw Hazel had it right so I called on her to give her a bit of confidence. I saw Kai had it right. But I was not surprised when he said it wrong. Robert was the same. He and Kai both knew the larger number but had probably guessed the smaller number instead of counting it. The work is totally inappropriate for Darren. He should not be in this group but I have to keep an eye on him. I waved Freddie away without a second thought — he knows the rules.

In this one brief instant of teaching Mrs D found herself sustaining the general pace of the action, monitoring individual responses, choosing one child to boost confidence, another to check understanding and yet another to attract attention. All the while she found it necessary to interpret their particular responses in the light of her knowledge of their typical behaviour. The wrong answer of one child was judged to be worth checking immediately whilst that of another was judged best left for later discussion.

In situations similar to those illustrated above the teachers perceived the demand for decisions to be incessant and exhausting. They admitted that on occasions they went into 'automatic pilot' and held periods of routine questioning in order to 'have a bit of a rest' or 'to give my brain a breather'. The challenge was not only in the sheer amount of information they had to process. It lay also in the frequency of difficult decisions which they felt had to be made. Some of these are illustrated in the following excerpts in which Mrs G set out to explore and discuss three-dimensional shapes. She had previously had a lot of examples of these in her mathematics area. The children had played with and talked about them. As a preliminary to further exploration, Mrs G decided to check the children's grasp of the notion of sphere. A globe had been passed round, its spherical properties emphasized and its technical name was mentioned several times.

Mrs G: Sphere. That's right. Now. Can anyone else tell me anything that is a sphere? Adam? [Who has his hand up].
Adam: Square.

At this point Mrs G recalled, 'I was ever so surprised he said that. I nearly fell off the chair. He is so keen and so shy. I thought, "how on earth am I going to cope with this without putting him down?"' She decided to ignore his answer and to get Adam to feel the globe and describe it as a sphere. She then turned her attention to another child. Seconds later, in giving a further example of sphere, Ben offered 'half the world'. Mrs G later noted:

> I thought 'oh dear!' I cannot get into all that now. If he had said it later when we had got the main idea established I could have developed it using plasticene. As it was, lots of the children were just beginning to get the idea and I thought it would be too confusing. I suspected Ben would have benefited from some extension but I decided to pass over his suggestion for the sake of the rest of them.

There followed a period of acceptable responses for sphere shaped objects, and then Samantha suggested 'a circle'. Mrs G thought: '"Help!" This is all getting to be more difficult than I had anticipated. I was not prepared to have the group sit and wait whilst we cut a circle out to make the concrete contrast. They would have lost interest.'

In this brief space of time Mrs G had perceived that she had had to resolve a number of difficult problems and had done so by making rather messy compromises. She had seen conflicts between a concern for individual and group needs, between intellectual challenge and children's personalities and between the management of materials and of time. Mrs G's experience was typical of the teachers' reactions to the few exploratory sessions observed. Once the teacher engaged in relatively free discussion, the pupils' responses became unpredictable. Levels of uncertainty in the teachers' decisions rose dramatically. Their reflections revealed a fine awareness of their children — not only as mathematicians but as humans with sensitivities as well as academic strengths and weaknesses. Also revealed was an awareness of the social context and the need to maintain the pace of interactions to sustain group interest. What the teachers' reflections did not reveal was an awareness of the increasing mismatch between their thinking and their behaviour. For the less routine and familiar was the children's behaviour, the more fixated became the teachers' actions.

The teachers frequently rushed to interpret children's responses and to correct errors. They did not pause to check or test their own interpretations, let alone the child's responses. Under the press of events the teachers tended to close options, become repetitive, allow repetitive responses from the children and generally reduce the proceedings to a routine.

In these respects the teachers were behaving like most humans under potential information overload. Information processing techniques are brought to bear to make as much information processing as possible into a routine so that attention can be focused on some central issue. This process of routinization may be understood by reference to learning to drive a car. Initially every manipulation requires total attention. Gradually one learns to steer and change gear at the same time. Ultimately one manages to drive through traffic whilst maintaining conversations. This is accomplished by making as much as possible of the performance into a routine. Unfortunately making matters routine is the complete antithesis of a searching discussion. The implication is that erudite mathematical discussions may be beyond the limits of human information processing in the press of classroom events.

Teachers' Implicit Theories

The press of events might not be the sole reason for the teachers' responses under these circumstances. From the range of events happening at any moment the teachers make their selections. We may presume that their choices are based on preoccupations that the teachers bring with them to the situation. In this research the teachers did not spell out their reasons for selecting what they focused their attention on. We are, therefore, left to speculate as to what ideas constrain the teachers' selective attention.

One powerful idea seemed to be that in their teachers' eyes children are extremely vulnerable both intellectually and socially. Teachers frequently mentioned the need to sustain children's confidence. As a general principle this is unimpeachable. But at the level of practice, at the first sign of a hesitant glance, the teachers tended to jump in and help. The price of confidence can be dependence. It is a price often paid in the maelstrom of classroom exchanges.

The teachers also evidently had the idea that children can be very easily misled into false notions. The teachers recognized the fertility of ideas proferred by their children but passed them over for the sake of some of the members in the group who were judged to be not ready to deal with them.

The third major theme which appeared to focus the teachers' attention was the need to get on with the curriculum in recognition of the scarcity of time. The teachers observed that they had a commitment to their colleagues to cover the school mathematics programme which generally took the form of a commercial scheme. In this context discussions always seemed to be pushed along at a brisk pace by the teacher.

It is easy to point out that each of these considerations completely defeats the object of sharing notions through discussions. A discussion necessarily involves the challenge of ideas and could hardly be sustained towards this end if the chairperson feels that the preservation of confidence carries a higher priority. Equally, discussions present opportunities for sharing and testing ideas amongst participants with a broad range of experience. There would be little point in discussing ideas only with those who were 'ready'. Whilst teachers have only a finite amount of time at their disposal, they do have considerable autonomy in the manner of its use. Discussions with a guillotine hanging over the chairperson's head are unlikely to be searching.

Whilst these arguments may have considerable logical merit in showing that the ideas which appeared to preoccupy the teachers sounded the death knell for discussion in principle (and, as it turned out, in practice), they are likely to have as much impact in a classroom as readings from Clausewitz would have on a soldier in no-man's land. For the fact is that the teacher's selection mechanisms are all highly adaptive to classroom life. Dispirited or confused children spell management disaster. An uncovered syllabus spells official and parental disapproval. Whatever the logic of the teachers' preoccupations as they go through the motions of conducting mathematical discussions, the reality is they must manage and be seen to manage their classrooms and to cover the syllabus. Their preoccupations are highly adaptive to meeting these demands under the circumstances in which they work.

Summary and Conclusions

Discussions are generally considered to be an important part of a child's mathematical experience and particularly so in respect of the development of skills associated with reflection and application. Despite the exhortation of decades, fruitful discussions are rarely seen in primary classrooms. This is so even in the classrooms of teachers who recognize and endorse their value. When discussions are seen, they are mainly teacher dominated, brief and quickly routinized.

It has been suggested that the sheer information load a teacher must process in order to manage her class and conduct a discussion makes the successful appearance of such work an unlikely prospect. Whilst there have been reported sightings of this rare event, its existence remains to be confirmed.

What are the implications of this argument? Certainly there are few implications for the teacher. It would be futile to write out advice to the effect that teachers should try harder. Nor will teachers be convinced by demonstrations of discussions given on a one-off basis by a lavishly prepared and equipped virtuoso and involving half a dozen children. Even when these extravaganzas are impressive as performances, they say little about what the children learned.

The implications of this chapter arise in the main for those who dictate how classrooms are resourced both intellectually and materially. If discussions are to be standard practice, then quite clearly the time will have to be made to conduct them in an undistracted fashion. This entails reducing drastically the breadth of the curriculum that teachers feel obliged to rush over. Secondly, it entails having more adults in classrooms to ease the teachers' management concerns. These are the very minimum requirements.

References

Bennett, S.N., Desforges, C., Cockburn, A. and Wilkinson, B. (1984) *The Quality of Pupil Learning Experience*, London, Lawrence Erlbaum Associates.

Cockcroft, W.H. (1982) *Mathematics Counts*, London, HMSO.

Desforges, C. and Cockburn, A.D. (1987) *Understanding the Mathematics Teacher: A Study of Primary School Practice*, Basingstoke, Falmer Press.

Easley, J. and Easley, E. (1983), *What's There to Talk About in Arithmetic*, Paper presented to the Annual Conference of AERA.

THE CONSTRUCTIVIST PERSPECTIVE

In the last few years a new theoretical perspective has greatly illuminated the processes by which children learn mathematics. This is constructivism: the view that children construct their own knowledge of mathematics over a period of time in their own, unique ways, building on their pre-existing knowledge. The emphasis in this view is on children's mental activity — the active construction of meaning — on the basis of the multiplicity of experiences and social interactions they have, including those at school. Since the received view is that there is a unique, objective conceptual structure to mathematics, this is quite a challenging perspective.

Among psychologists the constructivist view of learning is not new. Much of constructivist thought originates with Piaget. George Kelly's (1955) personal construct theory includes many of the central notions. Neville Bennett and Charles Desforges' (1984) research on the quality of pupil learning experiences is based on a constructivist theory of learning (due to Donald Norman). However, only in the last few years has constructivism pervaded thinking and research on the teaching of mathematics, where it is proving very fruitful. Significant work based on this perspective is being carried out in the USA, by leading researchers including Paul Cobb, Ernst von Glasersfeld and Les Steffe (see, for example, Cobb and Steffe, 1983; von Glasersfeld, 1987; Steffe *et al.*, 1983).

Constructivism in its most powerful form is more than a theory of learning. It is also a theory of knowledge. Perhaps the leading exponent of this philosophical brand of constructivism is Ernst von Glasersfeld. In Britain one of the leading constructivist thinkers, John Mason, is also interested in its philosophical aspects. However, this does not prevent his thought from being germane to the practising mathematics teacher, as his chapter shows. Barbara Jaworski provides the second chapter, where she grapples with the significance of constructivism for the mathematics teacher in the classroom.

References

BENNETT, S.N., DESFORGES, C., COCKBURN, A. and WILKINSON, B. (1984) *The Quality of Pupil's Learning Experience*, London, Erlbaum.

COBB, P. and STEFFE, L.P. (1983) 'The Constructivist Researcher as Teacher and Model Builder', *Journal for Research in Mathematics Education*, 14, 2, pp. 83–94.

KELLY, G.A. (1955) *The Psychology of Personal Constructs*, New York, Norton.

STEFFE, L.P., von GLASERSFELD, E., RICHARDS, J. and COBB, P. (1983) *Children's Counting Types: Philosophy, Theory, and Application*, New York, Praeger.

von GLASERSFELD, E. (1987) 'Learning as a Constructive Activity', in JANVIER, C. (Ed.) *Problems of Representation in the Teaching and Learning of Mathematics*, Hillsdale, N.J., Erlbaum.

13 *Teaching (Pupils to Make Sense) and Assessing (the Sense They Make)*

John Mason

The purpose of this chapter is to make some concrete proposals about assessment in mathematics. The question is what to assess, and how? Before answering that question, it is necessary to establish a perspective in which the suggestions are based. This leads me to begin with the question of 'making sense', and what this means when teaching. Only then is it reasonable to address the question of assessment.

My basic premise is two-fold: first, that pupils try to make sense of the world; and second, that they do so by assembling fragments of their experience into some sort of story. I arrived at this perspective as a result of marking scripts, and spending a large amount of time listening to what students had to offer at the end of a period of concerted work on a topic. I found to my surprise that even after working through the most carefully constructed exercises students remained mostly inarticulate, and their attempts to account for what they were doing were highly fragmented.

I have used two techniques in my research. At the end of a session I invite pupils to contribute to a pupil list of technical terms involved in the topic, and then to try to spin some sort of story using these technical terms which describes what they have been doing in the session. I call this *reconstruction* because pupils are explicitly invited to reconstruct their own story, rather than to 'learn' mine. The second technique involves setting up a relaxed and informal evening session in which I pose questions of a mathematical nature for discussion, and then *listen* to what the pupils bring from their recent 'learning'. The effect of these sessions has been to bring into question what it was that I and the students think we have been concerned with when engaged in traditional classroom activity. In particular the question comes up for me again and again: what is going on inside their heads?

Figure 1. Multiplication Tasks

Counting in twos

There are 2 wheels on the bicycle.

1 On 4 bicycles there are 2 + 2 + 2 + 2 = 8 wheels.

Instead of ADDING equal groups you can MULTIPLY.

Write

2 4 groups of 2 = 8

3 4 × 2 = 8

4 4 multiplied by 2 = 8

5 4 times 2 = 8

6 How many wheels are there on 7 bicycles? 14

7 How many wheels are there on a 6 bicycles b 10 bicycles? 20 12

What Are Pupils Attending to?

The accompanying extract came home from school in the hands of 7-year-old Lydia. The section shown in Figure 1 is the first of three, the others having a tricycle and a car respectively, but with the questions otherwise identical. She asked me to help her because 'I don't know what to do.' I asked her to start reading to me. She got to 2 + 2 + 2 + 2 and said, 'Is it eight?' I replied, 'What do you think?' She said in a tentative tone of voice, 'eight'. She skipped over the 'Instead ...' and went on to '4 groups of 2'.

'What does that mean?' she asked.
'What does 4 groups of 2 mean?' I asked her.
'Eight?'

I pointed to the bicycle and asked her what that was doing there. She didn't know. She went on to 4 × 2, then asked, 'What is "multiplied"?' I watched her carry on, using her fingers to do 7 bicycles, and in response to my raised eyebrows she did it again. The last bicycle question she did quickly. 'Is that all there is?' She set to, head down, pencil tightly gripped. She worked through the bicycles, the tricycles and the cars. Each question was tackled in turn.

What did she make of the task? What did the author intend her to make of it? I suspect that she was meant to see that the operation of multiplication is signalled and notated in a variety of ways, and that repeated addition is the same as multiplication. Did she? I doubt it. She looked at me in amazement when I asked her what it was about, as if to say, 'It's just a bunch of questions, Dad (you fool!).'

I conclude that it is not easy to point people in the direction of seeing the general in the particular, the sameness in different events, but that 'seeing the general in the particular' is one of the root processes in mathematics, and probably in every discipline. Indeed, different disciplines might be characterized by the features of situations which are attended to, and the ways in which generality is perceived in particularity.

The question — what is going on inside their heads? — is endemic to teaching. At its heart lies a tension arising from what Brousseau (1984) calls the didactic contract. This tension arises between pupils and teachers in the following way. The pupils know that the teacher is looking to them to behave in a particular way. The teacher wishes the pupils to behave in a particular way as a result of, or even as a manifestation of, their understanding of the concepts or the topic. The more explicit the teacher is about the specific behaviour being sought, the more readily the pupils can provide that sought after behaviour, but simply by producing the behaviour and not as a manifestation of their understanding. Tension arises because the pupils are seeking the behaviour and expect the teacher to be explicit about that behaviour, whereas the teacher is in the bind that the more explicit he is, the less effective the teaching. The question then arises as to how it might be possible to make positive use of the didactic tension rather than descending into a negative spiral in which the teacher is more and more explicit about the sought behaviour and the students more and more mechanical in their production of that behaviour. In reflecting on this question I have over the years made a number of self-evident but for me potent observations:

> *Observation 1:* I can't *do* the learning for my students.

Gattegno (1971) elaborates on this theme based on his memorable book title, *The Subordination of Teaching to Learning.* If I stop trying to do the learning for students, what are the implications? The ancient expression, 'there is no royal road to geometry', has applied more generally to learning mathematics for 2000 years, yet for 2000 years teachers have struggled to find the educator's stone!

> *Observation 2*: Students bring to class a rich experience of making sense in the past. Which of these powers do I particularly need to evoke in a given topic?

The expression, 'Starting where the students are', has unfortunately become a cliché, often meaning little more than not assuming the students know very much, whereas it could focus attention on helping pupils to use their undoubted powers effectively. Gattegno offers the challenging suggestion that 'only awareness is educable', and I take this assertion to encompass my question as a special case. My question then becomes, 'How do I evoke pupils' sense-making powers, and how do I help them work on their awareness?' Some people take Gattegno's expression to extremes, refusing to tell students anything on the grounds that it is useless unless the students discover it for themselves. Telling people facts is not in itself useless, indeed it is often essential, but the critical factor is how students go about making sense of what they are told as well as what they discover.

> *Observation 3:* Experience is fragmentary. We piece together bits of explanatory stories that we hear or construct, in an attempt to organize experience. We constantly probe our past experience to look for similar situations, and that similarity comprises the structure of our understanding.

As the Simon and Garfunkel song 'The Boxer' has it, 'A man hears what he wants to hear and disregards the rest.' 'Those that have ears . . .' We can only notice certain aspects of events, namely those we are attuned to or which stand out for us. We are not just solipsistic robots, however. We are constantly seeking resonance with ourselves and with others by expressing our stories and seeing what others make of them. The extent of that resonance provides confidence, and helps determine the company we keep.

> *Observation 4:* The result of making sense seems generally to consist of two elements: articulate stories which explain or account for a variety of situations; manipulative skills which are the subjects of examination questions.

The didactic tension leads to emphasis on manipulative skills, and on the conveyance-container metaphor for teaching. Students are 'given' skills, they try to 'get the point' of the lesson, they do or do not have a 'grip' on the concepts, and so on. It is interesting that in any discipline the development of techniques or skills for answering certain kinds of questions arises as the result of people observing a similarity or commonality to a whole range of questions, thus making it worthwhile to try to find a common solution. The common solution is then taught to students as a technique. The immense amount of construal which is required in order to reach a perception of commonality is not often shared with students, who are simply 'given' the skills without any reference to, or appreciation of, the original or underlying questions.

> *Observation 5:* Because as teachers we are engaged in transmitting culture to the younger generation, there is a tendency to move rapidly from a quick glimpse of an idea to succinct, manipulable, (symbolic) recordings — definitions, technical terms, major results.

As a culture we put great emphasis on the written expression of what people experience or think about. Some pupils have been recorded as saying that school is a series of events to be written about by pupils; thus writing becomes an aim or purpose of school rather than merely one manifestation of making sense.

Observation 6: The push to written records, and to manipulating symbols and technical terms, is due to a teacher's wish for pupils to automate procedures so that they can manifest the desired behaviour. These procedures are both particular, in the sense of being topic-based, and general or heuristic, in the sense of being thinking skills which are used throughout a particular discipline.

Reflection on these observations suggests that between *seeing*, in the sense of a vague and fuzzy glimpse, and *saying*, in the sense of striving for succinct verbal expression, and between manipulating examples and formulating articulate stories, lies the domain of mental imagery as source for and agent of the act of verbalizing (see Mason, 1986b, for elaboration). Furthermore, although the act of trying to express in words on paper is helpful for clarifying one's ideas, it is often difficult to write what cannot yet be said. Thus it makes sense to spend time trying to contact any mental imagery that is involved or associated with the topic, trying to express verbally to oneself and to colleagues, and only then trying to record or express ideas in pictures and words. Written records may go through many different drafts before becoming succinct and formal. The triad of *seeing, saying, recording* is a useful reminder that each contributes to the growth of understanding, and that a too rapid push to written records which omits the opportunity to modify, to try to express oneself verbally, may be so demanding as to block progress, and even to turn one against the particular topic.

The same sort of idea can be expressed from a different perspective by observing that it is generally considered good practice to invite pupils to carry out exercises, in the hope that this will literally 'exercise' their growing technical, manipulative skills, and give them access to the abstract ideas underlying the technique. There is, however, a good deal more hope than structure in such an approach. Whenever we get stuck on a problem or find some statement too abstract or general, we search in our experience for an example with which we are familiar. In other words, we turn to entities which are for us confidently manipulable, so that we can try to interpret the unfamiliar in a more familiar context. The act of manipulating, interpreting, exploring is more than simply doing exercises, because we are trying to get a sense of what the person is talking about.

In order to 'get a sense' we summon up various forms of imagery connected with past experience of similar situations. If we are given assistance by the speaker in the form of pauses, during which we can try out our examples, or ponder examples provided by the speaker, in which our attention is drawn to the features salient for the speaker, then by this suitable stressing and ignoring we can be assisted to re-experience the generality which we first heard. If we have the opportunity to try to express this generality for ourselves, then we begin to bring to articulation the vague sense that was germinating while working on the illustrative

Figure 2. The Process of Making Sense

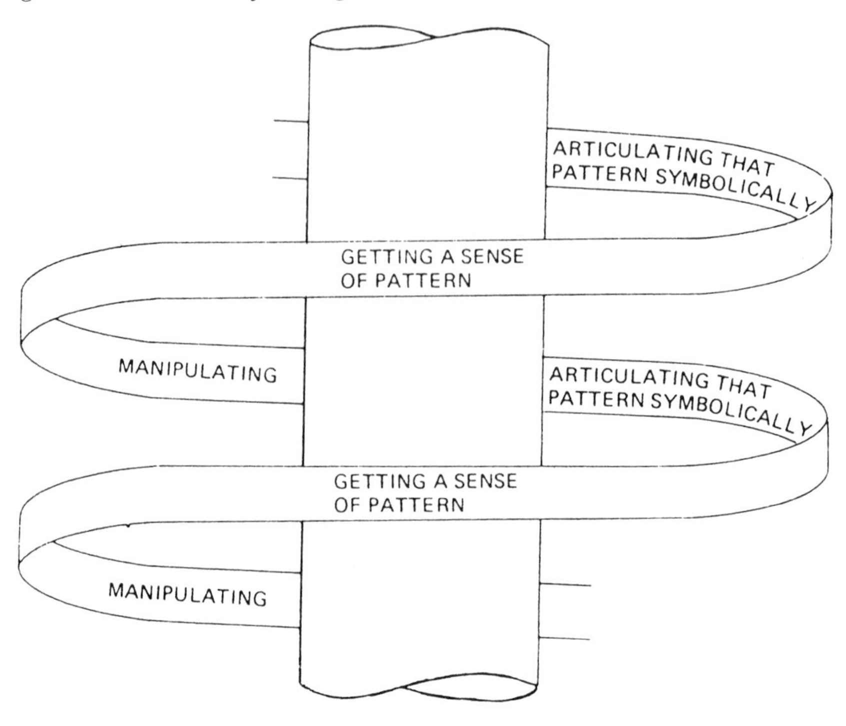

examples. Once we become familiar with, and confident with, manipulating the succinct expressions of the generality, we have new entities which might themselves be employed in later topics as illustrative examples of further ideas. It is convenient to display this process of making sense, of construing, as a helix in which the movement from confident manipulation through imagery and the processes of specializing and generalizing in order to 'get a sense of', to expressing and articulating the general principles in an increasingly succinct form, are seen as one complete turn around the helix (see Figure 2).

When we get into difficulty with something that someone is saying, we tend to move back down the helix through our images and through sufficient turns so that we find something confident that we can use as an example. We then retrace steps up the helix to try to re-express the generality for ourselves.

Teaching and Assessing

With these observations and remarks we can now address the title of this chapter. What does it mean to teach by helping pupils to make sense, and how might we go about assessing the sense that they make? The word 'teaching' is a curious one, because in some sense the one thing that students do *not* need is to know how to make sense of the world around them. Having survived to the age they have reached, having learned to speak, to walk and to run, they have illustrated all the powers necessary for making sense of the world. However, it seems that over the centuries individual disciplines have developed specific techniques for efficiently making sense of questions which are orientated in that discipline. Therefore, one thing a teacher can do is to work explicitly with pupils on the question of how you make sense in the particular discipline — in my case mathematics. Thus I need to evoke imagery, to learn to be explicit and precise about the imagery which I have inside me and from which I speak, to be explicit about the processes and methods of specializing and generalizing which pertain to my discipline, and to work explicitly with students on exposing and weaving together into stories the fragments of their experience which they recall.

Such remarks are rather general and unstructured, and so the rest of this chapter is concerned with offering a structure both to help students to construe in a discipline and to help teachers to assess that construal so that students at all levels of experience and ability can demonstrate what they *can* do rather than be penalized for what they cannot.

Before going into detail, it is important to acknowledge the fantastic tensions which are present in this approach. As a teacher there are a myriad of details and aspects to which I must attend. Is it not asking too much of pupils to be able to weave together articulate stories for complex ideas when in the past we have struggled simply to get a majority of pupils to be able to manipulate a few entities and to carry out a modicum of techniques in response to examination questions? No matter what ideals I carry as a teacher, when I find myself in a classroom with pupils in a particular state, is it not asking too much that they engage in the sorts of activities which are and will be suggested? In Mason (1986a, 1986c) there are elaborations of these and other tensions, and of ways to work on those tensions. As a teacher I can only work with the energy which is present in the pupils. If I am required to work on topics which are not yet of direct and immediate concern to them, then I must harness their energies and evoke their interest.

Activity of Teaching

The activity of teaching can be viewed from a four-fold perspective in which the current state of teacher and pupils, the aims of both, the resources available and the tasks embarked upon strive for balance and appropriateness.

What would I like students to achieve at the end of a topic or course?

Figure 2. The Process of Making Sense

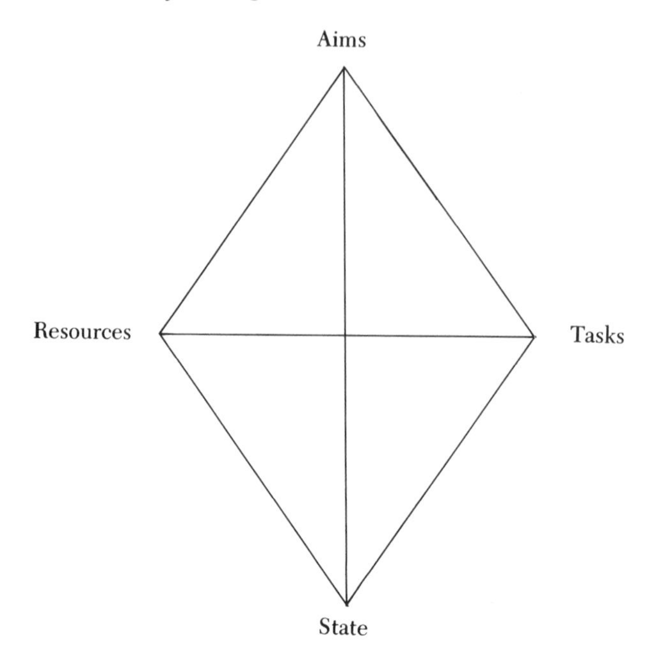

I would like students to have 'seen' connections, to have experienced some sort of integration or crystallization of disparate experiences which are subsumed under some general concept. I would like them to have 'gained a sense of' some coherence of a topic and how the techniques, technical terms and 'facts' fit together. I would like them to 'be articulate' about the meaning of various technical terms, of how standard and novel examples illustrate the ideas of the topic (see Michener, 1978, for elaboration); I would like them to be articulate about their own story of what the topic is about and what sort of questions it answers or deals with. I would like them to get to the point of employing succinct articulations confidently in the future as components or examples or tools in new topics.

These aims are teacher aims. Does 'subordination of teaching to learning' and 'starting where the pupils are' mean that my aims should be the same as the students' aims? I suggest not. Students quite naturally often wish to minimize their effort, and many are reluctant to stand out from their peer group. The didactic tension comes into play. But the aims I have outlined here are aims connected with making sense, not with showing off or with making extra effort. Consequently, bearing in mind my initial assumption about students wanting to make sense of the world, my aims and students' aims are at least confluent, if not identical.

The wishes outlined contain automated skills, general impressions and articulate stories, experience and familiarity with examples, and more generally with the effects of specializing and generalizing in this particular topic area. The extent to which pupils succeed in all of these aims will depend not only on the teaching style adopted, but on a host of factors including:

predisposition/interest/involvement in problematic questions at the heart of the topic;
peer group attitude to learning;
teacher attitude to learning and teaching, interest, commitment, etc.;
facility with assumed automated skills;
the extent to which pupils' *own* powers are evoked and employed in the teaching and learning;
the extent to which pupils share the teacher's goals.

These are some of the factors which make up the didactic situation (Brousseau, 1984). With so many influences it is clear that there is no royal road to learning or to teaching. Suggestions, such as reconstruction and listening, pausing, attending to the spiral through manipulating, getting a sense, articulating, attending to the back and forth flow between particular and general, are merely fragments of an ethos or *Weltanschauung*. They are devices intended to promote a perspective, and do not comprise a 'method'. The words are the results of attempts to draw distinctions which teachers have found useful, because in the midst of an event, suddenly becoming aware of the distinction reminds them that there may be some aspects they may have neglected or some alternative ways of engaging pupils and evoking their powers. In other words, Gattegno's assertion holds equally for students *and* teachers: 'Only awareness is educable.'

Since assessment is currently such a major concern for pupils, teachers and educators, it seems sensible to face the tide, and deal directly with how one might assess the sense that pupils make of lessons. This will in turn indicate a perspective which informs classroom practice so as to better prepare pupils for assessment.

Assessment

Taking the view expressed in the *National Criteria* (SEC, 1986) and the Cockcroft Report (1982) that assessment should provide pupils with an opportunity to show what they *can* do, and not to hide what they *cannot*, I suggest that assessment might sensibly look for:

evidence of what a pupil can *do*, at a functional, behavioural, technique level;
evidence of facility at various levels of explanatory coherence, showing a sense of, and where appropriate an articulateness about, particular concepts;

evidence of carrying out various mathematical thinking processes such as resorting to particular cases, seeking and expressing general patterns, convincing oneself and others;
evidence of participating constructively in group work and discussions, both as participant and as leader.

In the language of Skemp (1971) these aspects might be described succinctly as:

instrumental understanding of content;
relational understanding of content;
participation in process;
participation in social roles.

Consider a topic such as density, the Norman invasion, solving triangles, *Macbeth*, or any other that comes to mind. I submit that the nature and purpose of formal education is to facilitate movement of attention to and fro between particularities and generalities. In other words, it is to become aware of, and articulate about, patterns or generalities which encompass a variety of contexts and situations (Mason, 1984b). (Note this same movement in the use of particular examples like density and *Macbeth* in order to indicate the general.) Successful pupils can move from the particular to the general, and from the general to the particular. Pupils who are process-aware (but not necessarily articulate about it) quite naturally evoke both movements *automatically* as appropriate. In some situations pupils can operate only *reflexively* in the sense that they can employ both movements appropriately when reminded, but not always without being reminded. In some situations pupils can only *react* to explicit suggestions. They need specific help in invoking fundamental thinking processes which they have already used in order to learn to walk and talk, but which for some reasons are not employed in the particular lesson. The triad of *reactive*, *reflexive* and *automatic* helps some teachers to recognize differences in pupil responses that might otherwise have been overlooked, and thus enables them to extend pupils appropriately.

Movement from explicit and detailed work employing thinking processes (and thus developing and refining them), to reflexive triggering off processes by a word or gesture, to automatic invocation, is subtle. It can be assisted by the use of explicit vocabulary, but words can also become superficial jargon. Mason (1984a, 1986c) elaborates the tension between words as superficial jargon and as precise technical terms, and suggests techniques for developing and maintaining richness and meaning for didactic frameworks.

The six levels of performance shown in Figure 3 are based on an analysis of the particular-general movement, together with a distinction between being able to give an *account* of what a topic is about, and being able to *account for* various features or anomalies that appear in the topic.

Figure 3. Six Levels of Mathematical Process

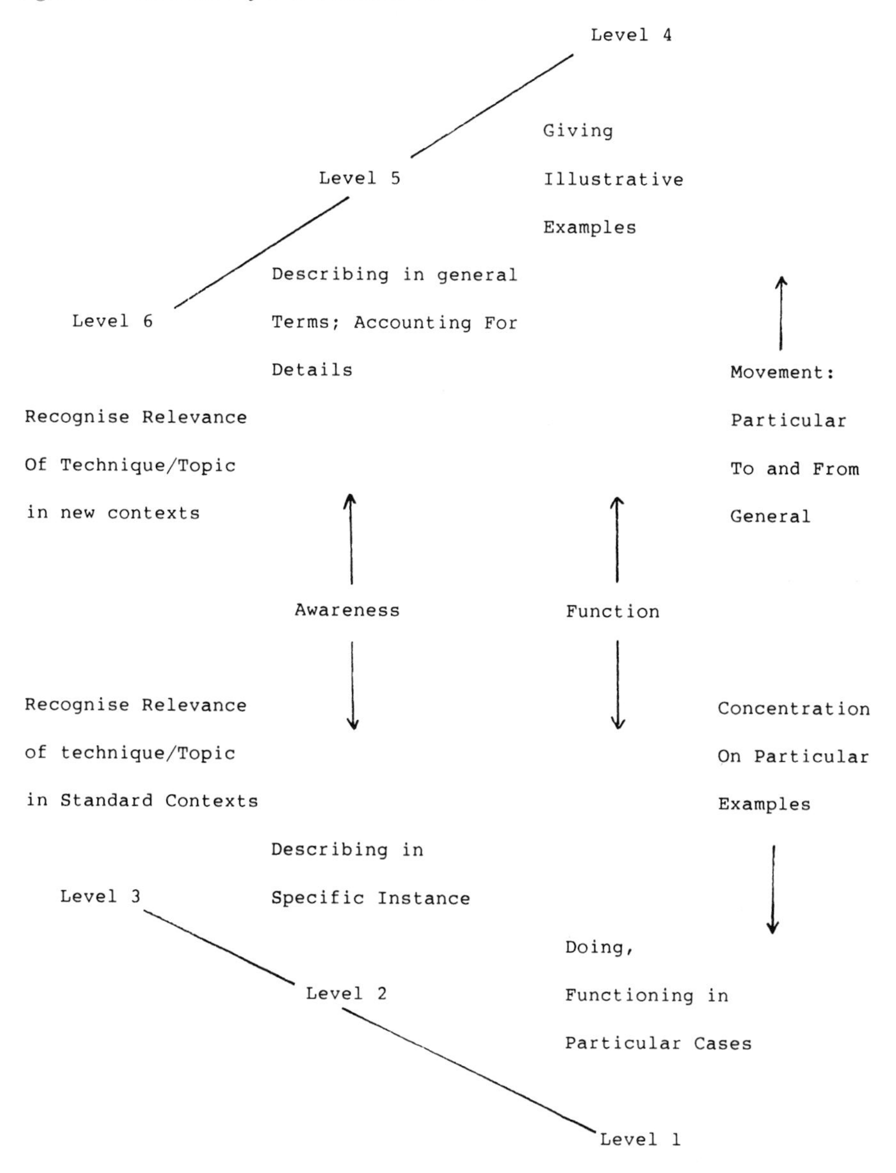

The six levels provide *both* a basis for designing assessment *and* a technique for helping pupils make sense of a topic for themselves — in short, to construe and verify their own meanings. An overall picture of the six levels is given in Figure 3, which can usefully be read from right to left, as a flow from the functional to the perceptive, from left to right as an unfolding of the essence into the functional, or as levels developing clockwise from bottom right round to top right. Elaboration with examples follows the diagram.

Most examinations test facility at level 1 and level 2 with respect to techniques like solving equations. Rarely are pupils called upon to show what they can do at higher levels. Similarly, pupils are frequently put to work on sets of exercises which are intended to develop facility at level 1 or 2. Rarely are they encouraged to give their own account *of* a topic, or to account *for* how topics fit together. Yet it is in the act of explaining to others ('you only really learn when you have to teach it') that most people really make a topic their own by constructing their *own* story of how all the details fit together.

Levels 1 to 3: Giving an Account of (Describing)

Level 1 Doing specific calculations, functioning with certain practical apparatus (e.g., add fractions of a particular type; make measurements; read tabular data; find a solution to a given pair of equations; find the reflection of a point in a line ...).

Recalling specific aspects of a topic, and specific technical terms (e.g., fractions can be added, multiplied, compared; equations sometimes have solutions and sometimes not ...).

Level 2 Giving an account of how a technique is carried out on an example in own words; describing several contexts in which it is relevant (e.g., you multiply these together and add those ...; you measure perpendicularly here ...; fractions arise as parts or shares of a whole ...).

Giving a coherent account of the main points of a topic in relation to a specific example (e.g., fractions can be compared by subtracting or by dividing ...).

Giving a coherent account of what a group did, in specific terms (e.g., we tried this and this, we noticed that ...).

Level 3 Recognizing relevance of technique or topic/idea in standard contexts (e.g., if two-thirds of a team have flu ... — recognize fractions; two kilos of coffee and one kilo of tea cost ... — recognize simultaneous equations).

Levels 4 to 6: Accounting for (Explaining)

Level 4 Giving illustrative examples (standard and own) of generalizations drawn from topic, or of relationships between relevant ideas (e.g., the simplest denominator is not always the product

— give an example; sometimes simultaneous equations have no solutions — give an example).

Identifying what particular examples have in common and how they illustrate aspects of the technique or topic (e.g., what does $5/6 + 3/8 = 23/24$ illustrate about adding fractions?; when this point is reflected in this line, and its image reflected in the same line, you get back to the starting point — what general rule is being illustrated?).

Level 5 Describing in general terms how a technique is carried out; to account for anomalies, special cases, particular aspects of the technique (e.g., to add two fractions you . . .; simultaneous equations with no solution arise because . . .; a triangle reflected in a line cannot be translated and rotated back to its original position — why?).

Level 6 Recognizing relevance of technique or topic in new contexts.

Connecting topic coherently with other mathematical topics (e.g., fractions are one way to get hold of certain kinds of numbers . . .; reflections are examples of transformations . . .).

The six levels are intended to suggest ways of structuring tasks for pupils which will provide impetus and opportunity for them to make ideas their own, as well as providing a format for assessment of content. The proposed format includes assessment of the major mathematical processes — specializing, generalizing and convincing. Other processes such as the use of, and switching between, representations will arise in the context of particular mathematical topics. Problem-solving and investigational work could be assessed using the same format, in which pupils are called upon to report (orally or in written form) on their own work and the work of their group at the various levels.

It is important to bear in mind constantly what the final assessment record might look like. It is intended to be positive, to state what the pupil has achieved, to give pupils something short-term to aim for, to provide reward and thus a boost to confidence. Any recorded evidence of completion of some task will have to be relevant and clearly understood by agencies who might later see the certificate — employers, colleges, etc. Assessment records will have to be simple enough for such agencies to comprehend, yet complex enough not to reduce twelve years of schooling to a few numbers. Performance on a single day under unusual circumstances is a peculiar way to sum up someone's potential. Giving pupils many opportunities to demonstrate competence is fairer to all concerned.

In Mason (1984a) it is suggested that the brick-by-brick construction metaphor for scientific epistemology is wildly inappropriate in (mathematics) education, that a more appropriate metaphor might be a forest, which is constantly changing, yet also unchanging; that Plato, and all teachers before and after, have had to face the same basic didactic tensions; that each generation struggles to express their questions and solutions in idioms appropriate to the times. This chapter illustrates the point, for its roots can be found in Sankya philosophy, Pythagorean qualities of numbers, Plato and Dewey — to name only a few. Pedigree is, I believe, much less important than vividness and resonance in the current didactic context.

References

Brousseau, G. (1984) 'The Crucial Role of the Didactic Contract in the Analysis and Construction of Situations in Teaching and Learning Mathematics', in H-G. Steiner (Ed.), *The Theory of Mathematics Education*, Occasional Paper 54, Bielefeld, IDM.

Cockcroft, W.H. (Ed.) (1982) *Mathematics Counts*, London, HMSO.

Gattegno, C. (1971) *What We Owe Children: The Subordination of Teaching to Learning*, London, Routledge and Kegan Paul.

Mason, J. (1984a) 'Towards One Possible Discipline for Mathematics Education', in H-G. Steiner (Ed.), *Theory of Mathematics Education*, Occasional Paper 54, Bielefeld, IDM.

Mason, J. (1984b) 'What Do We Really Want Students to Learn?', *Teaching at a Distance*, 25, 1984, pp. 4–11.

Mason, J. (1986a) 'Tensions', *Mathematics Teaching*, 114, pp. 28–31.

Mason, J. (1986b) 'I Is for Imagery and Imagination', *The Investigator*, 9, pp. 8–9.

Mason, J. (1986c) 'Challenge for Change: To Stay Alive Every Teacher Must Become a Researcher', *Mathematics Teaching: Challenges for Change*, AAMT 11th biennial Conference proceedings, pp. 9–30.

Michener, E. (1978) 'Understanding Understanding Mathematics', *Cognitive Science* 2, pp. 361–83.

Secondary Examinations Council (1986) *GCSE National Criteria*, London, DES.

Skemp, R.R. (1971) *The Psychology of Learning Mathematics*, Harmondsworth, Penguin.

14 *Mathematics Teaching: Belief and Practice*

Barbara Jaworski

I have been a mathematics teacher for many years, and have recently spent much time observing lessons and talking with other teachers about their teaching. This chapter is about some of my observations and related thinking.

A Teaching Reality

I had made plans to observe a lesson of a particular teacher one Friday morning, and then to talk with him afterwards during his 'free' period. The lesson was immediately after break so I arrived before break and walked over to the staffroom with him for a cup of coffee. On the way over a number of pupils stopped to ask him about the skiing trip that they were all taking over half-term, and he told them to meet him during the lunch hour to talk about it.

As we dodged through the crowded corridors, he raked his hair and said that he felt terrible. It had been a parents' evening the night before and he had not gone home until 10pm. He said that he had talked to so many parents that it was hard to remember what he had been saying by the end of the evening. One of his comments was: 'You know, if only we could come in [to school] an hour later, or something, the next morning — so that you could just get an extra bit of sleep — it wouldn't be too bad. As it is you just feel under constant pressure.' As we drank coffee, he suddenly swore as he found on the noticeboard that he had been asked to cover a class for the lesson in which we were supposed to talk. He said: 'I have three free periods a week, and this is the third time this week that I've had to cover.' The deputy head walked past at that point, and he challenged her. Her reply was that she sympathized, but they happened to have twelve staff away that day, and so what could she do?

As we walked back to the maths area at the end of break the conversation was about how with the best will in the world you could not find all the time that was needed for preparation of lessons, assessment of work and exciting teaching!

The Effectiveness of Teaching

When going into schools to observe their lessons, a number of teachers have apologized to me with words such as, 'I'm sorry. I'm afraid that its only KMP today.' The implication has been that I should be looking for something more exciting than KMP, or SMP, or whatever scheme or text was in use; that resorting to a text or scheme was something to be regretted. These teachers have admitted that they value more the lessons into which they are able to put the most energy in terms of planning and originality, but that it is impossible to be exciting and original for thirty-six lessons a week, and a buffer in the form of a text or scheme is essential to them. Moreover, the pupils very often like the schemes. They provide a comfortable environment for pupils in many respects. All of this raises questions about what it means to teach, and, in particular, what is *effective* teaching?

I can easily get a sense of the effectiveness of a can of defrosting foam by noticing how well it disperses the ice on my windscreen. As a teacher, to get a sense of the effectiveness of my teaching I have somehow to be able to judge how well it affects my pupils. To put a measure on this I have first to define the affecting that I consider valuable.

I believe that in every lesson a teacher asks, either implicitly or explicitly, 'What sense are the pupils making of the mathematics that they encounter in this lesson?' This question is often very difficult to answer, and it is possibly in finding ways of answering it that a measure of effectiveness might be found for any teacher.

I observed a class of 11/12-year-olds, working on area and perimeter of rectangles. The exercise in their textbook gave them the values of two of four variables — length, width, area, perimeter — and pupils had to work out the other two. Two girls asked me a question about the resulting area of 0.9, when they multiplied a length of 1.8 by a width of 0.5. As one girl put it, 'We can't see why it's 0.9. 0.9 is about the same as 0.5, not much bigger. We expected the answer to be bigger.' In our subsequent conversation they drew a rectangle and talked about the sizes of the sides, what area they would expect, and the result of multiplying the two numbers. I had the impression that this was the first time that they had really thought of the question in terms of area, rather than just in terms of multiplying numbers.

I knew that a previous lesson had been spent on an activity called 'Half-and-Half'. I could see the results of it up on the wall; some very impressive two-colour squares in which pupils had imaginatively split up the area into two halves and shaded half the area in each colour. The teacher had told me excitedly about how pupils had constructed and justified the construction of their squares. I asked the two girls which of the squares were theirs and how they had constructed them, and they showed me what they had done. I was impressed both with their creativity and with their explanation of why the result was half and half. They went on to tell me about a homework problem, which the teacher had set, in which someone started with 100m of fencing and wanted to enclose the maximum possible area within it. They had some ideas as to how to tackle the problem, but had not got very far with it yet.

As I thought of the three separate activities, I wondered what links the pupils were making between them, if any. I asked the two girls what they thought today's lesson had to do with the work on half-and-half, and with the homework problem. They told me that 'they're all to do with area, aren't they', and I wondered what else I might have expected!

I mentioned it to the teacher afterwards. I wondered, even if the pupils were articulate enough to express what they felt about it, whether the sense they were making, the links that they saw, would have coincided with the expectations of the teacher, or if they would have surprised or disappointed him. We wondered together whether asking the pupils frequently to try to explain any links that they saw might eventually lead to a more explicit making of such links. He decided to ask the class to write down a few sentences to say what they thought the topics had been about, and how they were connected, and then said wryly, 'That's hard isn't it? I'm not even sure that I could do it myself!'

This experience led me to think about what, as a teacher, I actually want to know when I ask, 'What sense are pupils making of this?' and how I can find out. I recalled a lesson which I taught myself some time ago at the end of which I asked pupils to write a sentence or two on what they thought the lesson had been about. The replies included:

1 We drew a shape and described it to our partner who drew it any way he thought right.
2 This lesson was about discussing and drawing your thoughts and trying to use other people's descriptions.

The replies were almost equally split between those pupils who described the activity in terms of drawing shapes (as in 1), and those who talked about the communication (as in 2). I remember feeling particularly pleased that some pupils had actually perceived the communication aspect, although they may have found it difficult to put into words. However, I was not surprised that many pupils had seen the activity mainly in terms of what they had done physically. In the case of the area activities the pupils' responses to the teacher's request were mainly factual about the particular activities, and conveyed little sense of pupils' general awareness of area.

There seemed to be a parallel here between reaching generality in terms of seeing mathematical rules from a number of particular cases, as in a textbook exercise for example, and having a general sense of a mathematical topic from a number of activities in which one has engaged. Mason and Pimm (1984) discuss the meaning of generic and general in such contexts. It seems an important objective of mathematics teaching that pupils should become aware of such generality, but how the teacher might monitor this awareness is not clear.

The Teacher's Role

A teacher's judgment of the effectiveness of a lesson is bound up both in what actually occurred and in how this is reconciled with the teacher's own beliefs about teaching and learning. It is often true that what seems ideally desirable is difficult to achieve in practice and the teacher has to struggle with implementation. However, some of the struggles which I have perceived recently have been as much to do with belief itself as with its implementation. If the struggle is with belief, then it makes the implementation even more difficult.

The following extract is from a conversation which I had with a teacher after a particular lesson on tessellation:

> Teacher: '... my aim was to follow one of the questions that was asked at the end of last lesson, which was "Why are some tessellating and some not?", because some people got to the stage where they saw hexagons tessellating and the quadrilaterals, but they found pentagons didn't, nor did octagons ... and the other aim was to work on children explaining more fully when they were discussing things. I'm not sure whether I achieved the first one or not....'
>
> '... I'm not sure that it worked exactly as I'd hoped it would work or that they actually focused on the angles meeting at a point as I'd hoped they might ...'
>
> '... They kept referring to the fact that if they were able to make the shapes into quadrilaterals or rectangles that they would be able to tessellate the shapes. But yet they weren't all convinced that quadrilaterals tessellated. That was the thing I wanted them to go on to....'

This teacher was struggling with the tension between wanting the pupils to explore their own ideas and come up with their own explanations, and wanting them to perceive certain mathematical results which she thought important. One option open to her was to explain her ideas to the pupils, but for the moment she was rejecting this.

The effectiveness of teacher explanation is measured in some sense by how well the pupils understand the explanation, and for many pupils a good teacher is one whose explanations are understandable. One perspective of teaching and learning mathematics is of a transmission process where mathematical knowledge exists and may be conveyed by the teacher to the learner. The teacher's responsibility is to give a clear articulate exposition of the mathematical knowledge, and the assumption is that having heard it the recipients then have the knowledge and ought to be able to say it clearly themselves, or give other evidence of understanding. Exercises are designed for pupils so that they can rehearse the ideas and provide this evidence for the teacher.

An obvious but nevertheless important consideration is that *the teacher cannot do the learning for any pupil.* Even the transmission process depends for its success on pupils constructing their own images of what the teacher has said, and trying to make sense of them. This trying to make sense is a complex process of relating what is heard to previous experience and making links which tie in the new ideas to established understanding or belief. von Glasersfeld (1984) describes this in terms of the 'fitting' a new experience into existing experience. We seek to explain what we encounter in terms of our existing belief, and understanding consists not of comprehending an absolute reality but of establishing such a 'fit'.

The word 'construal' might replace the phrase 'trying to make sense'. Construal takes place for everyone at all times in which they are aware and alert. Even though a pupil is not attending to the teacher's words, she is still construing. Her construal may involve a growing awareness that she is missing whatever the teacher is saying, and some establishing of an attitude to this, maybe of panic, of unconcern or of resistance to authority. The teacher engaged in transmission hopes that pupils' construal is of mathematics, and of the particular mathematics which the teacher has in mind. However, individual construal, as described above, implies that every pupil will have made a different sense of what was said according to their past experience and current thinking. So the teacher may have no assurance that all pupils believe exactly what she had in mind for them to believe.

One way for the teacher to find out what pupils are thinking is to get them to try to talk about it, to the teacher or to each other. If they talk to each other, the teacher cannot listen to everyone at once, but she can listen in to different groups and get an overall sense of what is being said. In one classroom, after a particular activity in which pupils had described the fitting of three red and three white Cuisenaire rods to a blue rod in different arrangements, the teacher asked pupils to talk in pairs about whether they thought the different arrangements were important. As she listened to two boys talking, it appeared that they were talking about football. One said to the other, 'If its QPR 3 and Bristol Rovers 3, it doesn't matter which order the goals were scored in, does it?' The boy's ability to translate the situation into an example that made sense for him gave the teacher an insight into his construal.

Pupils do not always find it easy to put mathematical ideas into words, which is not surprising since adults, and even mathematics teachers, often have difficulty themselves. If a teacher wants pupils to talk about mathematics, then she has to provide opportunities for them to develop this skill. Activities which are designed to require negotiation between pupils can provide this opportunity in a fairly natural way; for example, the activity where one pupil had to draw a shape in secret, then describe it to a partner who had to draw the shape which he understood from the explanation. When the second drawing did not match the first, the pupils were asked to discuss what would have improved the explanation. It was interesting to see how the discussions progressed from recriminations from the explainer, about the inability of the responder to understand what was said, to a realization that the onus for clarity was on the explainer as much as on the responder. The activity was done first in pairs, but then a number of pairs volunteered to perform for the whole class and the class commented on the quality of the explanations which they heard. The teacher listening was able to get a sense of the pupils' understanding as the negotiation progressed. The role of the teacher in this activity was mostly that of facilitator and listener, but the opportunities for learning about pupils' understanding were greater than in a lesson in which the teacher had done most of the talking.

An investigative approach to teaching and learning involves providing opportunities for pupils to express and explore ideas for themselves, and encourages pupils to ask their own questions and follow their own lines of inquiry. It requires confidence and flexibility on the part of the teacher and a willingness to explore whatever comes up. It does not prevent the teacher from posing questions which alert pupils to ideas which the teacher considers to be important, but it does involve the teacher in being prepared for a wider consideration of ideas than an expository style might allow.

The daughter of one of my colleagues had talked with her father about triangles on the surface of a sphere. She approached her maths teacher the next day with a challenge which I paraphrase: 'You know that you said that angles of a triangle add up to 180 degrees?' The teacher replied, 'Yes?'. 'Well, what if the triangle is on a sphere? The angles don't add up to 180 degrees then, do they?' The teacher was reported to have replied tersely that he had not been talking about triangles on spheres and that what he had told her had been quite correct for plane triangles. It is possible that the teacher felt both embarrassed and threatened by the challenge and that this conditioned his response. Being more welcoming of questions from pupils and more open to alternatives might avoid such embarrassment, but this is not always easy. One teacher, whose lessons I had enjoyed because her classroom always seemed to be a place of high energy and lots of pupil ideas and initiative, talked to me about her experience when she had to teach calculus to her fifth year O-level group.

> I've never taught calculus before, and I'm not very confident about it. I prepared myself thoroughly by reading up first in a number of textbooks, but I couldn't bring myself to make it open-ended. I taught it very straight from the blackboard and didn't really invite questions. When Elizabeth asked a question that I didn't know the answer to, I nearly panicked. But I was able to invite the rest of the class to comment on the question, and Janet said something that I realized was right and so I took it from there and it was OK.

How Can the Teacher Control What Pupils Learn?

It seems an inevitable step to come to the conclusion that the teacher cannot hope to prescribe what pupils learn, except in a very narrow sense; that having to work to a prescribed syllabus contradicts reason, as do the traditional forms of examination by which the syllabus is assessed. A compromise exists in practice, and individual teachers have to establish what this compromise means for them. One teacher is currently working on a balance between encouraging pupils to generate their own ideas and ways of working and exercising his responsibility as a teacher to help them in their development of successful strategies. 'I listened to what they were saying, and it was clear to me that they were in difficulties because they couldn't organize what they had found out so far. So I said, "well if I were doing this I would . . .".'

He had set up groups of four in which the two pairs were given different problems on which they were expected to work. Although the problems were different, the pairs were asked to talk to each other occasionally about what they were doing and thinking. One group of four decided to ignore this instruction and start together on the same problem. When they were bogged down in masses of data with no conclusions, the teacher came across to help. They expressed their frustration as, 'Clearly you have some result in your head which you want us to find, but we can't find it.' To ease their frustration he decided to give them some hints, but pointed out that had they followed his instructions, the other problem might have provided some of the insights which they required. He said afterwards that there are sometimes occasions when he does want pupils to find out a result which he has in his head, and one of his responsibilities is to find ways of making this happen. Sometimes it is appropriate to tell them, at other times he needs to find other ways. He is actively working on this problem as he teaches his classes from day to day.

A group of researchers at the University of Grenoble in France is actively looking into what questions and activities will bring pupils up against particular mathematical ideas. This research could be of considerable use to teachers who are struggling with such issues.

The pressure on teachers is unlikely to change. It can only be bearable if seen in the context of pupils' success. As teachers, the way in which we use time and resources is indicative of what we most value. We need to be asking continually about the sense which pupils are making in our classrooms. No major action is called for, just an awareness that pupils are trying to make sense in their terms and teaching has to make room for this. Every teacher can be a researcher, and every lesson can provide opportunities to find out more about how we can help pupils to learn.

References

MASON, J. and PIMM, D. (1984) 'Generic Examples: Seeing the General in the Particular', *Educational Studies in Mathematics*, 15, pp. 277–89.

VON GLASERSFELD, E. (1984) 'An Introduction to Radical Constructivism', in P. WATZLAWICK (Ed.), *The Invented Reality*, New York, W.W. Naughton and Co.

Part III

The Social Context

Mathematics teachers have long been aware of the importance society attaches to their subject. All school leavers are expected to be numerate, and an examination pass in mathematics is a required entry qualification in many areas of employment and for most courses of higher education. This importance has been reflected in a series of official reports on the teaching of mathematics from Board of Education (1909, 1912) to Cockcroft (1982). However, only in recent years has the relationship between society and mathematics been explored beyond the level of exhortations to improve. Following Griffiths and Howson (1974), there has been an increasing number of studies of the broader aspects of the social context of mathematics teaching. The Committee of Inquiry chaired by Sir Wilfrid Cockcroft commissioned both a survey of curriculum development and research in the teaching of mathematics (Howson, 1983) and a survey of research on the social context of mathematics teaching (Bishop and Nickson, 1983). Leading British researchers in the field hold regular symposia on research into social perspectives of mathematics education, convened by Steve Lerman and Marilyn Nickson. Increasing recognition, both British and international, led the Sixth International Congress on Mathematical Education, (Budapest, 25 July to 3 August 1988) to devote a full day to 'Mathematics, Education and Society'.

With all this interest the question arises: what issues does a consideration of the social context of mathematics teaching raise? Although opinions vary, the following are key questions:

What societal forces determine the mathematics curriculum?

What are the aims of teaching mathematics? What is the role of mathematics in general education?

What social and political values are implicit in the mathematics curriculum?

Is mathematics teaching meeting the needs of all its clientele, including the needs of the less academic and the lower social classes?

Mathematics and gender: why are girls disadvantaged? What can be done to remedy this?
Multicultural mathematics: how should mathematics education reflect our multicultural society?
The profession of teaching: what are the consequences of its structure?
How autonomous is the mathematics teaching? What factors play a role in determining autonomy?
What is the social status of mathematics and the mathematics teacher?
Teacher education: is it a qualitative or a quantitative problem?
Is mathematical knowledge a social construction, and if so what is the basis for mathematical truth?

Excluding the last, these are some of the immediate questions that arise concerning the relationship of mathematics, education and society. The last question is less immediate and more philosophical, concerning as it does the social context of mathematics, not the teaching of mathematics. Yet it may well be this last question that holds the key to answering the others. The relationship between mathematics and society holds the seeds of their relationships with education.

Traditionally mathematics has been viewed as the paradigm of certain knowledge. Euclid erected a magnificent logical structure 2300 years ago, which has been taken as the paradigm for establishing truth and certainty. Newton copied it in his *Principia*, and Spinoza in his *Ethics*, to strengthen their claims of uncovering truth. The nineteenth century showed us that Euclid's system is flawed, and modern mathematicians have created amazing systems in the hope of providing a certain foundation for mathematics. But all in vain. Every attempt to establish the certainty of mathematics failed. Furthermore, mathematicians established results showing that secure foundations for mathematics are impossible (for example, Godel's theorem on the incompleteness of proof, and Tarski's proof of the undefinability of truth). In the last fifty years there has been growing recognition that mathematics is not a body of objective truth, but a social construct — as human a creation as any other body of knowledge (see, for example, Lakatos, 1976). Although carefully constructed, mathematics has no more claim to infallibility than any other human knowledge.

Once mathematics is seen for what it is — as much a cultural product as any other — a number of myths about mathematics can no longer be sustained. The myth of the objectivity of mathematics is exploded — scholars are forced to admit that it does not hold. Mathematics is not a monolith — the unique objective structure is also nothing but a myth. For there are different specialisms starting with different axioms; there are incompatible treatments of the same specialism, for example, classical and constructive analysis; those studying applied mathematics, statistics, groups, sets and logic, analysis, geometry, topology, number theory, and so on — are engaged in very different activities — do they really have a common language?

To the extent that it exists, the uniformity of mathematical meanings amongst mathematicians, and a shared view of the structure of mathematics, result from an extended period of training in which students are indoctrinated with the 'standard' structure. This is achieved through common learning experiences and the use of key texts, such as Euclid's *Elements* in the past. Most students fall away during this process. Those that remain have successfully negotiated and internalized a substructure that fits with a subset of the official one — at least in part. Thus the shared, agreed structure of mathematics is an artefact, achieved through the institutionalization of the structure through community acceptance. All that can be claimed is that individuals' reconstructions 'fit' with the official structure of mathematics — which may be purely mythical. The work of constructivist researchers suggests that even in a fixed cultural context of schooling the order of acquisition of mathematics is unpredictable and uncontrollable — and that unique personal structures result.

The key outcome of this philosophizing is that mathematics must be seen as embedded in a cultural context. The view that mathematics somehow exists apart from everyday human affairs is a dangerous myth that cannot be sustained. It is dangerous because in addition to being philosophically unsound, it has damaging results in education. If mathematics is a body of infallible, objective knowledge, then mathematics bears no social responsibility. The underachievement of sectors of the population, such as girls; the sense of cultural alienation from mathematics felt by many groups of students; the relationship of mathematics to human affairs such as the transmission of social and political values: its role in the distribution of wealth and power; the mathematical practices of the shops, streets, homes, and so on — all of this is irrelevant to mathematics.

On the other hand, once it is admitted that mathematics is a living social construct, then the aims of teaching mathematics need to include the empowerment of learners to create their own mathematical knowledge; mathematics can be reshaped, at least in school, to give all groups more access to its concepts, and to the wealth and power its knowledge brings; the social contexts of the uses and practices of mathematics can no longer be legitimately pushed aside, the uses and implicit values of mathematics need to be squarely faced, and so on.

This second view of mathematics as a dynamically organized structure located in a social and cultural context identifies it as a problem-solving (and posing) activity. It is a process of inquiry and coming to know, a continually expanding field of human creation and invention, not a finished product. Such a dynamic problem-solving view of mathematics, in the mind of the teacher, has classroom consequences. In terms of the aims of teaching mathematics the most radical of these consequences are to facilitate confident problem-posing and solving; the active construction of understanding built on learners' own knowledge; and the exploration and autonomous pursuit of the learner's own interests.

Such an approach to the teaching of mathematics legitimates ethnomathematics — the naive, intuitive, pre-academic, culturally embedded conceptual structures and practices of mathematics. It also raises the importance of the study of the history of mathematics, not just as a token of the contribution of many cultures, but as a record of humankind's struggle — throughout time — to problematize situations and solve them mathematically — and to revise and improve previous solution attempts.

As this discussion and the previously framed questions show, a consideration of the social context of mathematics and mathematics teaching raises some important and controversial issues. Many now feel that the traditional status of mathematics and mathematics teaching as value-free and politically neutral can no longer be taken for granted. The social import of this is shown by the four issues treated in this part of the book: gender and mathematics; social and political values in mathematics teaching; multicultural and anti-racist mathematics teaching; and the profession of teaching. The above discussion of the nature of mathematical knowledge touches on all of these issues. They are all of deep social and political consequence: matters of importance for the education community, and for society in general. Some will mourn the loss of neutrality that the new, socially aware perspectives bring. Others see this as a sign that mathematics education is coming of age; it is beginning to take responsibility for the actions carried out in its name. Together the chapters show that now, more than ever, for those with the courage to probe beneath the surface of things mathematics teaching can be an exciting intellectual adventure.

References

Bishop, A.J. and Nickson, M. (1983) *A Review of Research in Mathematical Education Part B: Research on the Social Context of Mathematics Education*, Windsor, NFER-Nelson.

Board of Education (1909) *The Teaching of Geometry and Graphic Algebra (Circular 711)*, London, HMSO.

Board of Education (1912) *Special Report on the Teaching of Mathematics in the United Kingdom (Volumes 1 and 2)*, London, HMSO.

Cockcroft, W.H. (1982) *Mathematics Counts*, London, HMSO.

Griffiths, H.B. and Howson, A.G. (1974) *Mathematics: Society and Curricula*, Cambridge, Cambridge University Press.

Howson, A.G. (1983) *A Review of Research in Mathematical Education Part C: Curriculum Development and Research*, Windsor, NFER-Nelson.

Lakatos, I. (1976) *Proofs and Refutations*, Cambridge, Cambridge University Press.

GENDER

A key issue concerning the social context of mathematics teaching is the difference in mathematics achievement between men and women. For a decade evidence has been accumulating that women and girls come out of mathematics education disadvantaged relative to men and boys. Thus Hilary Shuard, in an appendix to the Cockcroft Report (1982), was able to document this disparity quite fully at the beginning of the 1980s. Given the key role of mathematics as a 'critical filter' in obtaining employment, this is a very important matter, not only because is it a source of inequity in our society, but also because it means that society is not fully utilizing the talents of half of its members.

Given the fact that girls on average achieve less well than boys in mathematics, the question arises as to what the cause might be. Are women's minds different from men's — somehow less mathematical — or do the differences arise from the environment — social conditioning? This is the notorious nature versus nurture dichotomy, which is so hard to settle empirically. However, there are different social pressures and expectations on women, so that social conditioning seems certain to be playing a part in this process, at the very least. Some researchers have conjectured that women are conditioned to be less confident in developing 'autonomous learning behaviours' and are less 'risk-taking', to account for poorer performance in mathematical problem-solving. The two chapters in this section have tried to dig even deeper in their search for causes, looking at girls' belief systems. Both of the authors, Leone Burton and Zelda Isaacson, are well known for their work in this area.

Reference

Cockcroft, W.H. (1982) *Mathematics Counts*, London, HMSO.

15 *Images of Mathematics*

Leone Burton

Recently I have been working with some top junior pupils who were inventing their own mathematical games. The children had chosen with whom they wished to work and had devised their game, made it, created the packaging to house it and then tested it within their own class and with younger pupils in the school. Finally, the class decided upon a set of criteria which made a game 'good', and applied their criteria to their own games. As a result, six games were chosen as the best. Five of the six had been made by groups of girls. I discussed this with some girls in the class.

Leone: Why do you think five of the six chosen games were made by girls?
Pupils: Oh, that's obvious — the best people in the class at maths are girls.
Leone: So, you are all good at maths?
Pupils: Oh, no, not us! Louise is though, Louise is brilliant at maths.

Louise, who according to her teacher is by far the most able child in the class, sat muttering deprecatingly about her own ability.

Leone: Well, are you all looking forward to the maths you will do at secondary school?
Pupils: No, definitely not.
Leone: Why not?
Pupils: Because it will be too hard.
Leone: Has it been hard this year?
Pupils: No.
Leone: When you were 8, did you expect it would be hard when you were 11?
Pupils: Yes.
Leone: But you say it hasn't been hard this year.
Pupils: That's because as it got harder, we got older.

Leone: But, at secondary school you will be older than you are now.

Louise: But everyone knows that boys get cleverer at maths as they get older, so it will get harder for us and easier for them.

Many of the issues which are currently the focus of research interest are present in this short exchange. The statement as a whole reflects the recorded disparity in mathematical performance between girls and boys as conventionally measured by results at formal examinations. Thus the Royal Society reported that, while 1984 entries and results for CSE were fairly evenly distributed between girls and boys at each grade, the distribution at grade A of the O-level examination was roughly 2:1 in favour of the boys, girls being 47.3 per cent of the entry. The results at grade A of the A-level examination were 3:1 in favour of the boys, girls being 30.3 per cent of the total entry (Royal Society, 1986). While the performance of girls on tests taken in the primary years tends to be comparable with boys, and Louise herself could accept that it was reasonable that she should out-perform the boys at age 11, she has articulated a misconception built upon actual bias in mathematical involvement and performance. Everyone might not 'know' that boys become cleverer at maths as they get older, but there is certainly a tendency for many people to behave as if boys' styles of learning, their interests and the link between mathematics and their future careers constitute the norm. The imbalance is then seen as a problem which is located in girls. For example, Elizabeth Fennema and Penny Peterson (1985) have defined 'autonomous learning behaviours' as necessary to 'do high level cognitive tasks in mathematics. [They] include working independently on high level tasks, persisting at such tasks, choosing to do and achieving success in such tasks.' Fennema and Peterson assert that 'females are more dependent than males', where dependency is defined as 'seeking physical contact, seeking proximity and attention, seeking praise and approval, and resisting separation.' But this value-laden definition of dependency is by no means necessarily the opposite of the independence cited above. For example, Louise and her friends had certainly worked independently on the invention, construction and testing of their game, they had persisted and they had achieved success. But they were also extremely articulate in the advice that they offered to some younger pupils who were about to begin a similar project. They said:

make sure you choose to work with friends with whom you will be able to sort out difficulties so that you don't waste time quarrelling;

make sure that you think ahead, and plan what you are going to do and why, before you get started, so that you don't lose time and get frustrated;

make sure that you pool all your ideas so that you can choose the best;

share out the jobs according to who wants and likes to do certain things. That way you will get the best possible game.

Here was a group of girls building upon their autonomous learning behaviours in a context of 'proximity' and 'approval' but where the autonomy was a function of group rather than individualized action. Further, the group was adamant that they had learnt together and through one another and that the experience had been a positive one for all. This group was not unique. The other groups of girls reported in the same terms.

Autonomous learning behaviours are probably as important to the learning of other disciplines as they are to mathematics, and certainly girls' success in the non-technical, non-scientific areas indicates that they do not lack these behaviours, where their acquisition and use is seen to be appropriate and consistent with their own and others' expectations. The percentage entry of girls for English at A-level in 1984 was 70.7 per cent, almost exactly the same as that of boys for mathematics (68.7 per cent) (Royal Society, 1986). In my view there is a sufficient amount of bias experienced through patterns of socialization over the period from birth to the end of formal education to explain divergence in subject preference and performance. It is always salutary to remember that the total number of pupils succeeding at mathematics is too low to warrant any complacency about competency of any pupil, girl or boy, whether at the individual level or in terms of social needs. When we focus on the situation which is highlighted by the discrepancy in performance between girls and boys, we are raising questions about the learning experiences in mathematics which are offered to all our pupils.

What image of mathematics do the pupils have? Let the fourth year juniors speak for themselves:

[Girl:] I like maths but as you go along it becomes harder and takes more time ... I'm not looking forward to going to my new school and doing maths there.

[Girls:] Some maths are easy and some maths hard, we find add, takeaway and times quite easy ... we prefer doing math investigation because we find it easier than normal maths.

[Girl:] Maths are sums like division, taking away and lots of other things as well ... when you come to do an assessment test there are always things we do not understand ... I am not looking forward to maths in senior school.

[Boy:] Maths is learning to add, subtract, divide and multiply.... Maths will help us in later life to get good jobs and to work well in our jobs. I don't really like maths, but it is necessary. I think Maths stands for Massive And Torturing Hard Sums.

[Boys:] Maths is an educational subject which enables you to learn about adding, dividing, subtracting and timesing, etc.

[Boys:] Maths is addition, subtraction, division and multiplication you need it when you go in sweet shops because if you buy a Mars bar you might not get the right amount of change back. You need it in cricket and football results. SMP involves maths as well. [textbook authors please note!]

These children had been working at mathematical projects for the majority of their time during the term preceding their writing these comments. Many aspects of the remarks are worrying. The girls refer to their feelings about maths and particularly their concerns about changing schools. But the boys are just as convinced that, despite being necessary, mathematics is not very pleasant! Despite a great deal of effort, all of mathematics is still just the four rules.

The image of mathematics as being difficult arises from the children's early contact with the subject. Far from developing a mathematical language to match and complement the demands that they are making upon it, children are introduced to a codified language and rules for usage which are distanced from their own concerns and largely irrelevant to them. As many researchers and authors have pointed out, children 'need to be capable not only of operating within the formal code, but also of making fluent translations between formal and concrete representations of the same problem' (Hughes, 1986). These translations between formality of coding and the contextualization of the concrete representations being recorded do not appear to present a barrier to girls when linking their language development with the skills of reading and writing. Nor do the girls appear to stumble over their 'dependency'. However, girls' facility with language is validated. They are not only expected to be successful with reading and writing, but their interest in stories, poetry, drama, singing, indeed anything connected with language, is seen as appropriate. At the same time teachers perceive as 'natural' the exploratory, constructional, 'tinkering', often disruptive behaviour of the boys. Such patterns of behaviour are often inadvertently reinforced by teachers through their very attempts to 'deal' with the boys. The more attention the behaviour attracts, the more likely it is to be confirmed. One is left wondering how much the stereotypes of compliant, submissive, undemanding girls and boisterous, noisy, attention-seeking boys are themselves self-fulfilling prophecies resulting from parent and teacher behaviour. The result of such confirmation can be that, arguing circuitously, the very behaviour is then identified as necessary to success in learning mathematics, and the negative descriptors are appended to the girls who are failing to learn in this style. Boys' attention-seeking behaviour which is enthusiastic, eager and confident is defined positively, despite being just as demanding and reward-dependent as the negatively defined girls' behaviour. The moral of this story is that it is extremely difficult to resist the effects of gender stereotyping even when framing researchable questions.

Louise's strong convictions about the mathematical potentialities of boys spring, therefore, from the social stereotyping of mathematics, science and technology as a male domain and the coincidence of power, both in the human and the technical sense, with maleness. The images in the media, in language, in advertizing reinforce the roles and behaviour of active, powerful, controlling men and attentive, submissive and attractive women. As Gilah Leder (1986) pointed out, 'The preoccupation with the negative rather than the positive aspects of success in articles depicting successful females is unfortunate, and may help to explain not only why certain goals have different values for males and females but also why fewer girls than boys concentrate on the still male perceived areas of mathematics and science.' In a society where scientific superiority is an integral part of power and control it is not surprising that the domain of mathematics, science and technology is attributed to men. A trivial, but pervasive, example was the recent advertizing campaign by the Action Bank, which spelt out the name of the bank using human figures to represent each of the letters. All the figures, except two, were male. The two were a bespectacled secretary standing ready with her typewriter awaiting instructions, and a gardener entwined around a flowering tree! Jacobsen (1985) recently pointed to differences in child-rearing practices, peer group expectations and social attitudes, all of which contribute to the image of mathematics as a male domain. He cited an incident at the Twenty-Fifth International Mathematical Olympiad in Prague in 1984 at which he was 'firmly told by the team leaders of two countries that the percentage of women competitors can never exceed 8 per cent, as this is due to biological reasons which we simply cannot alter.'

Child-centred education which has been so enthusiastically embraced especially by primary teachers can provide a framework within which the differential needs of certain categories of pupils are not considered. Many primary teachers say that they do not notice the sex of a pupil, they are too intent on the pupil as a person, and yet numerous research studies have demonstrated that teachers do treat children differentially according to their sex. For example, in making comments on children's work, they tend to criticize boys for presentation, girls for the quality of the mathematics (Dweck *et al.*, 1978). In class they tend to interact with boys much more frequently than with girls, and girls have many more days when they do not interact with the teacher at all. High-confidence boys interact at higher cognitive levels more frequently than do high-confidence girls (Reyes and Fennema, 1982). Walden and Walkerdine

(1986) have drawn attention to the phenomenon which they called rule following/challenging and how this relates to the procedural rules of mathematics and the behavioural rules of the classroom.

> Challenging the internal rules of the mathematical discourse, relating particularly to the teacher's authority as guardian of those rules is important in producing what the teachers describe as 'real understanding'. Such challenging requires considerable confidence because it necessitates the recognition that rules are to be simultaneously followed and challenged. That many girls do not have such confidence, nor would dare to make a challenge offers a different explanation of girls' mathematical development than one which relies on the naturalistic or immutable.

Autonomy of learning is the objective set by most teachers for the pupils in their charge. However, a classroom full of autonomous learners can be an extremely challenging and vulnerable place for a teacher to be. Organizational devices such as textbooks which control the curriculum and leave the teacher free to adminster are introduced in part to reduce that vulnerability. Behavioural rules which require girls and boys to line up separately, to be listed separately on the register, to dress distinctively, even to hang up their coats in different places, are all mechanisms which are used in schools to reduce the pressure of numbers and the potentiality for loss of control. There is increasing evidence that when the staff of a school focus on equal opportunities, looking at such overt mechanisms, as well as considering teacher and pupil behaviour in the classroom, the impact of sexism in the curriculum, the use of sex-biased examples in texts, materials and examinations, the sex of the role models within the school, and so on, the impact is measurable in the response of the pupils. As Stuart Smith (1986) reported: 'In recent years, there has been a remarkable improvement in the number of girls from the school who have been successful in the 'O' Level Maths examination. This improvement . . . appears to be closely related to a number of steps which have been taken to end the masculine image of Maths in school.' Equally, Hazel Taylor (1986) asks: 'how effective isolated strategies can be without the support of a general commitment to equal opportunities throughout the school?'

It is unreasonable to expect that teacher input or school environment can effect radical changes in the image of women and of mathematics in society. At the same time it is unreasonable to deny that, as a social institution, the school has considerable influence on the attitudes and behaviours of those within it. It is important to remember:

> such changes do not happen quickly but are cumulative and therefore respond to structural adjustments of the kind mentioned above;

changes in attitude demanded of pupils must be seen to be consistent with the attitudes of teachers;

stereotyping incidents should be discussed openly within the class and the time given to such discussion be seen as important;

criticism of materials or books does not imply throwing them away. Heightening sensitivity to sex stereotyping is effectively done through critical discussion of the materials or books and by raising the question of what would have been more acceptable. In holding such discussions, many primary teachers are horrified to discover the firmly held sex-stereotyped views of some of their pupils.

In conclusion, it might be worth reflecting on what you, the reader, would say to Louise, and in what ways you would ensure that what you said was reflected in the experiences that you provided for her in your classroom. Remember that next September Louise will enter her secondary school. Recall the anecdote of another girl who started secondary school in 1984:

> It was the start of the new school year in a secondary school. The boys and girls were lined up outside their first mathematics class. As the teacher supervised them filing in he said: 'Girls, sit at the back because mathematics is not such an important subject for you as it is for the boys.' (Open University, 1986)

Acknowledgment

I wish to thank the children and teachers of Hurst Junior School, Bexley, for their participation in the work reported here.

References

Dweck, C.S., Davidson, W., Nelson, S. and Enna, B. (1978) 'Sex Differences in Learned Helplessness, II: The Contingencies of Evaluative Feedback in the Classroom and III: An Experimental Analysis', *Developmental Psychology*, 14, pp. 268–76.

Fennema, E. and Peterson, P. (1985) 'Autonomous Learning Behaviour: A Possible Explanation of Gender-Related Differences in Mathematics', in L.C. Wilkinson and C.B. Marrett (Eds), *Gender Influences in Classroom Interaction*, Orlando, Fla, Academic Press.

Hughes, M. (1986) *Children and Number*, Oxford, Basil Blackwell.

Jacobsen, E. (1985) 'Reducing Differences in Mathematical Expectations between Boys and Girls', *Studies in Mathematics Education*, 4.

Leder, G. (1986) 'Mathematics Learning and Socialisation Processes', in L. Burton (Ed.), *Girls into Maths Can Go*, London, Holt, Rinehart and Winston.

Open University (1986) *Girls into Mathematics*, Cambridge, Cambridge University Press.

Reyes, L. and Fennema, E. (1982) 'Sex and Confidence Level Differences in Participation in Mathematics Classroom Processes', Paper presented at the annual meeting of the American Educational Research Association, New York.

Royal Society (1986) *Girls and Mathematics*, Report by the Joint Mathematical Educational Committee of The Royal Society and The Institute of Mathematics and Its Applications.

Smith, S. (1986) *Separate Tables? An Investigation into Single-sex Setting in Mathematics*, London, HMSO.

Taylor, H. (1986) 'Experience with a Primary School Implementing an Equal Opportunity Enquiry', in L. Burton (Ed.), *Girls into Maths Can Go*, London, Holt, Rinehart and Winston.

Walden, R. and Walkerdine, V. (1986) 'Characteristics, Views and Relationships in the Classroom', in L. Burton (Ed.), *Girls into Maths Can Go*, London, Holt, Rinehart and Winston.

16 *Of Course You* Could *Be an Engineer, Dear, But Wouldn't You* Rather *Be a Nurse or Teacher or Secretary?*

Zelda Isaacson

That fewer girls than boys engaged in mathematics at school and beyond, and that girls on average perform less well than boys, especially at the higher levels of achievement, are well documented (e.g., Burton, 1986; Isaacson, 1982; Royal Society, 1986). These facts are increasingly a cause of concern in the UK and indeed internationally.[1] In recent years the flow of literature on the 'gender and mathematics' issue has substantially increased, and much of this writing has been concerned with possible explanations of the phenomenon of female underachievement.

In this chapter I shall suggest two theoretical constructs, namely, coercive inducements and double conformity, which individually, and even more together, offer powerful explanations and far-reaching insights, and are therefore potentially of immense value to researchers in this field.

The notion of a coercive inducement is one which Helen Freeman and I developed some years ago. It seemed to us that to suggest that girls were prevented from taking up non-traditional roles in society — or 'not allowed' to engage in 'boys' subjects at school and higher education levels — was far from the felt experiences of most girls and women. Experientially most girls do not see themselves as forced into submissive roles, or coerced into taking up low status and often subservient jobs — or into studying child development rather than physics at age 14+. Rather, these female or feminine roles, jobs, school subjects, etc. are chosen by them, but they are 'chosen' because of a system of rewards and approvals which act as inducements and which are so powerful that they come to be a kind of coercion.

These patterns are clear if one observes young girls at play and at work in infant school classrooms.[2] Girls regularly choose to play in the Wendy House (it may officially be called the 'domestic play area', but children and teachers alike slip into the familiar name); they do not often choose to play with Lego or other constructional toys; they choose pretty clothes, 'Care Bears' or 'My Little Pony' for their birthday presents; 6- and 7-years-old girls often see their adult lives in terms of getting married and having children. If asked whether they will have a job when they grow up, even academically able girls name such things as nursing or teaching, or being a secretary or working in a sweet shop. It is crucially important, however, to acknowledge that the choices girls make day by day and hour by hour are, on the face of it, very attractive ones, and that the intrinsic and extrinsic rewards (e.g., personal satisfaction, the approval of significant others) accruing to girls for appropriately feminine behaviour (caring for others, helping others, building up relationships with others) are a further reinforcement of these patterns.

As Helen Freeman and I wrote in a draft paper when we were developing the notion of a coercive inducement a few years ago:

> It has been suggested that a submissive role is forced upon girls through punishment of non-conformist behaviour. It seems to us, however, that it would be closer to the truth to suggest that, rather than being coerced into 'feminine' behaviour, girls are induced by a system of rewards and approval to accept a more passive role.

In a recent paper (Isaacson, 1986) I expressed the idea in this way:

> There is a sense, I believe, in which many girls are persuaded to adopt typically female modes of behaviour and to choose stereotypically feminine occupations and life styles because the rewards for 'feminine' behaviour are too great to be refused, rather than because they are prevented from choosing others. . . .

In 1987, in a 'That's Life' programme on television, viewers were shown a small girl taking part with an adult male partner in a ballroom dancing display. The rewards she obtained for this were enormous — applause, lots of attention, being treated as 'grown-up' and being dressed up in a miniature version of the bespangled costumes of the adult women dancers. Very few children would be able to resist such delights. Certainly this little girl revelled in her role. The inducement to play the feminine role was so great that in a very real sense it could not be refused by that child and so was a form of coercion — an offer which could not be refused. All the little girls who watched that programme, too, were being exposed to this coercion. The message came through loud and clear: 'be a real woman, wear pretty, spangled clothes and dance on a strong man's arm, and you too will be applauded and feted.'

What has any of this to do with mathematics learning? Very simply, it is that the learning of mathematics cannot be divorced from the social context in which that learning takes place. Mathematics (and science and technology) carry strong 'male' images, partly because they are seen as 'hard' — not necessarily intellectually difficult, but hard as opposed to soft (or feminine, yielding, etc.). The male image of these subjects has a long history. This image is further reinforced by the fact that they have traditionally been dominated by men, and that their public concerns are typically masculine concerns, such as warfare, machinery and work aimed at subduing or controlling nature (see, for example, Easlea 1981, 1983). Girls and women often come to believe, therefore, that these 'masculine' subjects, and the jobs they lead to, are not for them. Worse, they often believe (at a subliminal level even if this is not made explicit) that if they engage in these activities they will put their valuable femininity (valuable because of the rewards it brings) at risk.

The effect on mathematics learning is cumulative. At option choice time girls are often reluctant to continue with study of the physical sciences and technical subjects because of the male image they carry, because of a relative lack of experience in these subjects which goes back to infant school days, and because they are making positive choices for 'feminine' subjects (human biology, needlecraft, art). At this stage of education we see mathematics becoming increasingly unpopular amongst girls. This is in part because in itself it carries an offputting male image, reinforced by boys in mixed-sex classrooms who claim the subject as their own. It is also because one of the most compelling reasons currently for pupils to learn mathematics — that it will be useful for other subjects studied and for future careers — is also absent for the girl who has already opted out of science and/or technology and has decided that that sort of job is not for her. There is research which suggests that pupils who study physics or technical subjects alongside O-level mathematics do better at mathematics than those who do not (Sharma and Meighan, 1980). This reinforces the hypothesis that pupils who study these subjects (1) practise their mathematical skills in another context, and (2) have strong motivation to learn mathematics. So, although mathematics is a compulsory subject for the vast majority of pupils to age 16 in the UK, adolescent girls often disengage themselves from the activities of the mathematics classroom. Although they are generally not permitted to drop out of mathematics altogether, they can and do 'vote with their feet' and gradually drop down — from near the top of a class to nearer the bottom or from a high set into a lower one. The observable outcome is that many girls do not achieve as much as might have been predicted for them at an earlier stage of their schooling.

I grant that the system of rewards and approval for feminine behaviour which I said acts as 'coercive inducements' would not normally be called coercion. The coerced person is usually understood to be doing what they are unwilling to do, because of fear of unpleasant consequences ('your money or your life'). The coerced person is thus normally understood to be unfree, whereas acting in response to an inducement is usually regarded as acting freely. I wish to argue, however, that this juxtaposition of the two concepts offers a way into understanding the mechanisms of female underachievement. Girls and women who 'choose' the path of conventional femininity are in one sense acting freely — they could have chosen otherwise — but in another sense are unfree. When the rewards for being a 'proper woman' are huge, while by not conforming one risks the loss of these rewards, then one's freedom is no more real than the freedom of a person living below the poverty line to take an expensive holiday abroad.

The notion of a coercive inducement on its own, I suggest, goes a long way to explaining why females are underrepresented in 'male' subjects and occupations. However, when combined with the second construct under consideration in this chapter, that is, double conformity, its explanatory force is greatly increased. I am indebted to Sara Delamont for this latter idea. In an essay entitled 'The Contradictions in Ladies' Education' (1978) she claimed that:

> The central theme which can be traced through the establishment of education for middle and upper class girls and women from the 1840s to the present day is double conformity. This double conformity — a double bind or catch 22 — concerns strict adherence on the part of both educators and educated to two sets of rigid standards: those of ladylike behaviour at all times and those of the dominant male cultural and educational system. (p. 140)

Double conformity expresses the dilemma of any person who is in a situation where they have to conform at the same time to two sets of standards or expectations, where these two sets are mutually inconsistent. This was the case for the pioneers of women's education in the nineteenth century. It is also the case for many women today who reject stereotypical career choices but then find themselves competing with men in a world where the rules have been made by men to fit in with the ways in which men are expected to behave.

Research carried out some years ago into people's views of what are the characteristics of a mentally healthy, mature, socially competent (1) adult, (2) man and (3) woman, is very revealing (Broverman *et al.*, 1970). The characteristics of a normal adult and a normal man match very closely, while those of a normal woman are quite different. It is not possible (according to these profiles, reflecting views held by both men and women) to be at the same time a normal woman and a normal adult! This

> . . . places women in the conflictual position of having to decide whether to exhibit those positive characteristics considered desirable for men and adults, and thus have their 'femininity' questioned, that is, be deviant in terms of being a woman; or to behave in the prescribed feminine manner, accept second-class adult status, and possibly live a lie to boot. (p. 6)

A woman working in a male dominated and male defined sphere finds herself continually faced with having to choose between acting in ways which are appropriate to her as a woman, and appropriate to her as, say, an engineer. For men in these jobs there is no such conflict, whereas for women the conflict is an inevitable part of the job and, indeed, of being a mature and responsible adult in a sex-stereotyped world.

The combined effect of coercive inducements and double conformity is to increase enormously the obstacles which women have to overcome when they try to make their way in male dominated and defined areas of study and work. Competence and confidence in mathematics play a part in many of these, and girls who opt out of mathematics, science and technology at school, because they do not wish to enter these fields, are responding to very strong influences. They can hardly be said to be making choices based only on talent, interest or inclination. Although not usually expressed in this way, many girls see the choice as between living their lives under the stress of double conformity and being continually in a 'conflictual position' or, alternatively, gaining the rewards for conventionally feminine choices and behaviour. These rewards may well be short-term and short-lived, but life beyond age 25 is not salient in the eyes of most girls.

When I consider the gender and mathematics issue with the aid of these explanatory constructs, I find myself ceasing to be puzzled by girls' underachievement in mathematics, but rather astonished that girls and women achieve as much as they do.

If changes are to be brought about, the loss of female mathematical talent abated, and greater equality of opportunity and genuine freedom of choice opened up, then we have to look both at and beyond school practice in ways which take account of these deep-rooted forces. We have to work simultaneously on a number of fronts. One of these is to change the climate within mathematics classrooms so that ways of working which girls find comfortable are welcomed. An example of this would be to develop classroom practices and types of classroom organization which discourage competitive behaviour (where the search for the right answer is dominant), and instead encourage cooperative, collaborative and exploratory behaviour where each person's contribution — as an individual or as a member of a group — is valued. I find hopeful in this respect the directions in which GCSE mathematics, properly applied in the classroom, could take us. Another is to look carefully at the content of the mathematics curriculum, and ensure that this reflects a broad range of human concerns, rather than being narrowly focused on traditionally male concerns only. These sorts of changes serve to reduce the level of conflict for girls. We have rightly, and at long last, gone beyond the days when it was believed that improving girls' participation in mathematics required that girls were to be changed!

In Holland, when an alternative mathematics curriculum was intro-

Figure 1. Influences on the Learning of Mathematics

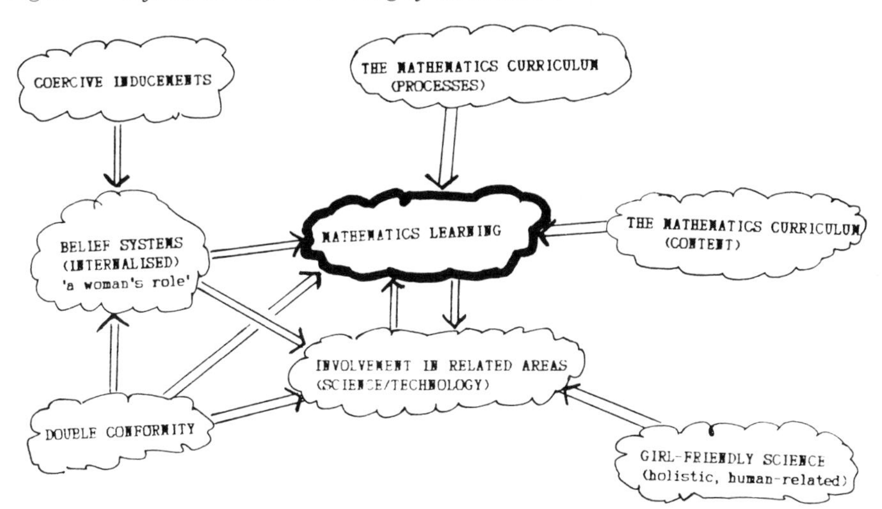

duced (Math A), with a higher 'social' content and broader-based applications which are more obviously and immediately relevant to pupils, the proportion of girls studying mathematics to age 18 greatly increased (Isaacson *et al.*, 1986). The Dutch experience suggests that changing the content of the curriculum alone can make a significant difference — how much improvement might we see if a number of the important variables were changed at once! Figure 1 offers a schematic view of the complex of interrelating variables which are key influences on mathematics learning.

Finally, I believe that work on gender and mathematics must go on, not only because of arguments derived from justice (women should not be discriminated against) or because of arguments derived from need (we cannot afford to neglect so much potential talent), important though both these are. A further and most compelling reason is that through work on this issue we begin to learn much which is of broad significance for the pedagogy of mathematics. Research and discussion on gender and mathematics are leading to a deeper understanding of the factors which influence mathematics learning in general, not just in females. This has been an unexpected bonus, and may in time prove to have been as valuable an outcome of the work as those originally intended — perhaps even the most valuable in the long term.

Notes

1 IOWME (International Organization of Woman and Mathematics Education) holds regular international meetings and publishes a biannual newsletter. In 1986 in

London a number of 'state-of-the-art' reports from around the world were presented, as well as information gleaned from the SIMS (Second International Mathematics Survey) data, all pointing to a continuing imbalance between female and male achievement in mathematics. In March 1987 the Dutch government funded a conference organized by Vrouwen and Wiskunde (Woman and Mathematics) on the theme 'Images of Mathematics'. In the UK the publication by the Royal Society of a report on *Girls and Mathematics* (1986) is a measure of the public recognition now given to this issue.

2 The author spent a term in an infants school in 1986, observing gender differences in children's play, style of dress, choice of companions and activities. She also explored with 6- and 7-year-old children their views on a number of questions through in-depth individual interviews.

References

BROVERMAN, I., BROVERMAN, D., CLARKSON, F., ROSENKRANTZ, P. and VOGEL, S. (1970) 'Sex-role Stereotypes and Clinical Judgments of Mental Health', *Journal of Consulting and Clinical Psychology*, 34, 1, pp. 1–7.

BURTON, L. (Ed.) (1986) *Girls into Maths Can Go*, London, Holt, Rinehart and Winston.

DELAMONT, S. (1978) 'The Contradictions in Ladies' Education', in S. DELAMONT and L. DUFFIN (Eds), *The Nineteenth Century Woman: Her Cultural and Physical World*, Beckenham, Croom Helm.

EASLEA, B. (1981) *Science and Sexual Oppression*, London, Weidenfeld and Nicholson.

EASLEA, B. (1983) *Fathering the Unthinkable*, London, Pluto Press.

ISAACSON, Z. (1982) 'Gender and Mathematics in England and Wales: A Review', *An International Review of Gender and Mathematics*, ERIC.

ISAACSON, Z. (1986) 'Freedom and Girls' Education: A Philosophical Discussion with Particular Reference to Mathematics', in L. BURTON (Ed.), *Girls into Maths Can Go*, London, Holt, Rinehart and Winston.

ISAACSON, Z., ROGERS, P. and DEKKER, T. (1986) 'Report on IOWME Discussion Group at PME 10', *IOWME Newsletter*, 2, 2, pp. 3–6.

ROYAL SOCIETY (1986) *Girls and Mathematics*, A Report by the Joint Mathematical Education Committee of the Royal Society and the Institute of Mathematics and its Applications, London, Royal Society.

SHARMA, S. and MEIGHAN, R. (1980) 'Schooling and Sex Roles: The Case of GCE 'O' Level Mathematics', *British Journal of Sociology of Education*, 1, 2, pp. 193–205.

SOCIAL AND POLITICAL VALUES

Although it has long been acknowledged that societal forces determine the mathematics curriculum, the view that there are social and political values implicit in it has been a controversial one. However, in the political climate of the late 1980s it is increasingly difficult to sustain the traditional view of mathematics and mathematics teaching as value-free and politically neutral. For example, the issue of whether mathematics in the classroom should foster cooperation or competition is now seen in some quarters as a political question.

Recent events have confirmed this: a consideration of some aspects of the social context of mathematics and its teaching has led to a great furore. The June 1986 CSE mathematics examination (London Regional Examination Board) presented candidates with data on world military spending and then asked:

> 'The money required to provide adequate food, water, education, health and housing for everyone in the world has been estimated at 17 billion a year' (*New Internationalist*, 1980).
>
> How many weeks of NATO + Warsaw Pact military spending would be enough to pay for this? (Show all your working)

In the days following the examination, newspapers carried headlines and subheadlines including:

> Sinister (*Sun*, 14 June 1986)
>
> Row as maths CSE examines arms spending (*The Guardian*, 14 June 1986)
>
> Examiners to vet maths paper for political bias (*The Guardian*, 14 June 1986)

> Question: What has arms spending to do with a maths exam? (*The Daily Mail*, 14 June 1986)

The fact that the mathematics question was placed in a political context — although not partisan in terms of East-West balance — was enough to trigger exaggerated reactions in the media (sinister?). What may be seen as politically unacceptable by some is that the question invites the student to compare the costs of armaments with those of meeting people's basic needs. Faced with the fact that world armament expenditure exceeds the cost of meeting all people of the world's needs several times over, it is difficult not to conclude that the arms race is immoral, especially when millions regularly perish from drought, starvation and curable diseases. But the question only presented basic facts, not these conclusions. There is a great divide between using controversy to provoke critical thought and exploiting education to promote propaganda. In the editor's opinion this example lies firmly in the former category.

This examination question provides an example of the most overt way in which mathematics can carry social and political overtones. That is through the context or situation to which a question refers. The first of the three papers in this section treats this dimension of values, but also considers the meaning of the hidden curriculum. The way mathematics is taught is also value-laden, for example, approaches can be democratic or dictatorial; mathematics can be taught as objective, infallible truths or as a way of knowing through which the learner can create personal knowledge; and so on. Another key aspect of the politics of mathematics education is the way it helps or hinders the distribution of power and wealth in society. This issue, which was already mentioned in the discussion of gender, is treated in all three of the papers in this section.

17 *Social and Political Values*

Paul Ernest

Social and political values have an impact on mathematics education in a number of ways. In the following I consider three of these areas of impact: the direct impact of societal values on the mathematics curriculum; the authoritarian-democratic continuum applied to mathematics education; and the isolation and fragmentation of mathematics.

The Impact of Societal Values

Societal values pervade the mathematics curriculum, both overtly and covertly. Consider the range of topics included in the mathematics curriculum. There are potent social reasons for including topics such as the following: social arithmetic, numeracy skills, measures, money, statistics, logic, ratio and proportion, variation and percentages, amongst others. There is no purely mathematical justification for the inclusion of percentages in the curriculum, for instance. There is no reason to distinguish hundredths as a particularly interesting denomination of fractions. Notations such as 33⅓% for ⅓ show that no mathematical simplification or unity is achieved. Rather percentages are taught for the sake of merchandizing and finance, as an introduction to the central role of commerce and banking in our society. The inclusion of this topic amounts to a tacit endorsement of the commercial aspects of our society. The point is not that this is either good or bad, but that this aspect of the curriculum is rarely discussed; there is too little awareness of the values underlying the curriculum, as was pointed out by Jenny Maxwell (MT111) and by Dawn Gill (MT114).

The mathematics curriculum also serves to endorse the values of technological 'progress'. The enthusiastic acceptance of various calculators, computers, interactive video players and other products of technology amounts to an endorsement of a system of values — the valuing of materialism and technological innovation for their own sake. Do we ever stop to question the significance of these toys?

Societal values are also manifested in the aims of the mathematics curriculum, aims which correspond to different interest groups in society. Three aims may be distinguished:

the *utilitarian aim* concerns the acquisition of utilitarian mathematical skills; the achievement of mathematical functionality;
the *personal development aim* concerns the contribution of mathematics education towards the growth, development and general education of the whole individual;
the *mathematical aim* concerns the transmission of mathematical knowledge; the communication of the academic discipline of mathematics to students.

These aims reflect the values of society as a whole (especially the government and employers), of humane and liberal thinkers and educationalists, of the mathematics community, respectively. The power of these groups over the mathematics curriculum is reflected in the priority given to these three aims. The mathematics curriculum is dominated by the mathematical aim, followed by the utilitarian aim with occasional gestures towards the personal development aim.

Should this order not be reversed? Personal development is vital, mathematical functionality is important and beyond this the acquisition of mathematics per se is of little general significance. Some steps towards the recognition of this have been made in Great Britain by the Cockcroft Report and the new GCSE examination system for 16-year-olds in their recommendation of 'bottom-up' mathematics syllabuses, based on the notions of success and functionality for all as opposed to the older 'top down' approaches dominated by the structure of mathematics. In addition, the increased emphasis on the processes of mathematics including problem-solving, investigations, project work and pupil discussion amounts to the recognition of the personal development aim of mathematics education.

For too long mathematics teaching has been dominated by narrow mathematical and functionality aims. A consequence is that mathematics is perceived as isolated from broader social and political issues, and is not used to develop critical thought. This in turn promotes the uncritical acceptance of social norms, or worse still, the irrational and destructive rejection of social institutions.

Authority versus Democracy

One of the major components of the hidden curriculum of mathematics teaching concerns the pupils' experience of authority. Values representing a position on the authoritarian-democratic continuum are communicated to pupils in many of their mathematics learning experiences. This is also

Figure 1: Endpoints on the authoritarian-democratic continuum

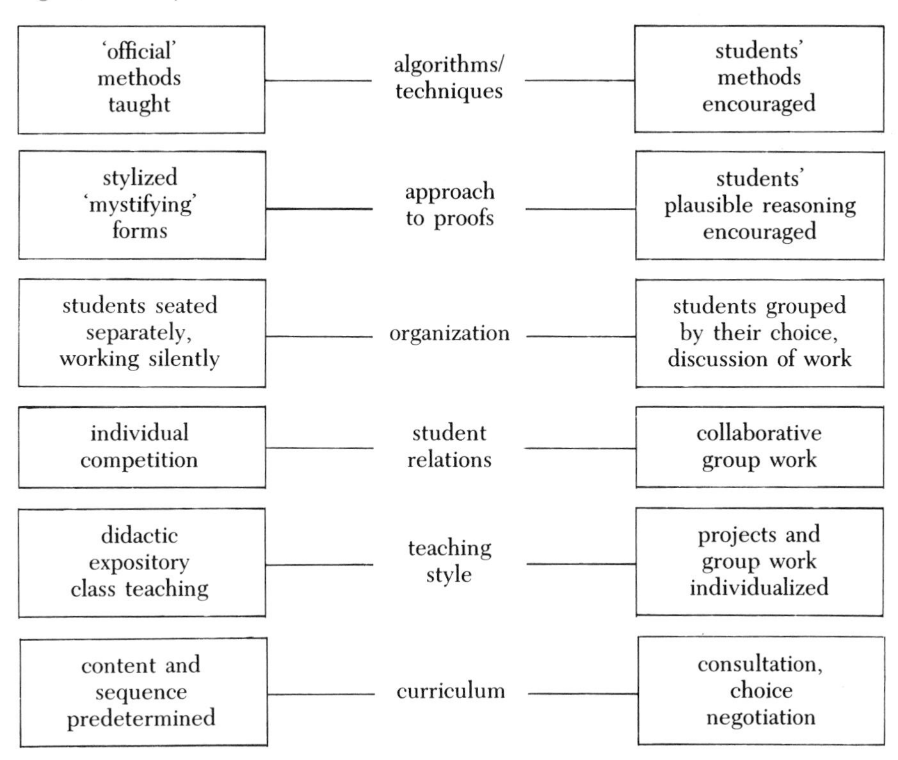

true in other areas of the curriculum, but is especially marked in mathematics.

Students' experiences of mathematics may be termed authoritarian when they are in a position inferior to and in submission to the teacher whose dictates and teaching must be followed and absorbed without question. In this extreme position students are fully dependent on their teacher for every aspect of their mathematics learning.

Students' experiences of mathematics learning may be termed democratic when the classroom atmosphere is one of relative freedom, where they are respected and respect each other as individuals, where they become increasingly independent of the teacher in their learning, and they are free to discuss and negotiate many aspects of the curriculum.

It is possible to construct a continuum from authoritarian to democratic modes for various aspects of school mathematics:

the ways the subject is presented (the status of definitions, the approach to proof, the attitude to techniques and algorithms);
the ways a student's work is dealt with (the forms of assessment used, how errors are handled, answers checked);
classroom management (seating, access to resources, the way students' tasks are selected, the sort of questions a teacher asks);

relationships which are permitted, encouraged or discouraged (between students, between students and teacher);
the curriculum (how it is chosen, the way different parts are appoached, its orientation — whether it is directed towards the students' experience or interests).

Readers will be able to extend these suggestions. It is then instructive to consider this multitude of dimensions and to identify the extremes of the continuum. For instance, Figure 1 suggests authoritarian extremes on the left, democratic ones on the right. The location of these aspects of the mathematics curriculum on the continuum determines the extent to which a student's experience of the hidden curriculum in mathematics is authoritarian or democratic. Unless a substantial number of these features is experienced at the democratic end of the continuum we will not be preparing students to participate in our modern democratic societies as confident and independent critical thinkers.

Fragmentation and Isolation in Mathematics

A further set of values which may be uncovered within the mathematics curriculum, and indeed beyond in the broader mathematics community, concerns the unity of mathematics and its connections beyond its own boundaries. Mathematics is too often taught as a sequence of unconnected skills and topics. Mathematics is too often taught in isolation from the world of its applications. Mathematicians are often regarded as inward looking, clustering in groups to talk about their shared imaginary worlds just as we see in young adventure game enthusiasts.

During an official survey of 10 per cent of English secondary schools, fully 78 per cent of the mathematics departments sampled were criticized (41 per cent strongly criticized) for their lack of cooperation with other subject departments.[1] This illustrates the isolationism of mathematics and mathematicians. But there is a deeper sense in which mathematics is isolationist. In the last decade or two there has been a growing concern about the social responsibilities of science. Radical scientists have come to realize that the fruits of technology and science are not ethically and morally neutral in view of their potentially awesome power. There has not been a parallel growth of awareness of the social responsibilities of mathematics. The received view of mathematics is that it is neutral with regard to social and political issues. This is an isolationist view, which divorces mathematics from its social and political context.

Four ways in which mathematics is linked with this broader context are worth considering: the military aspect of the mathematical sciences, the monoethnicity and sexism of mathematics, the dehumanization of mathematics, the use of mathematical abstraction to dehumanize social issues.

The military associations of mathematics are venerable; consider only the study of projectiles and ballistics. More recently Turing's hypothetical computer was made into reality in World War II to decode the Enigma ciphers. Subsequently the government of the United States of America financed the development of our present miniaturized computers as on-board guidance systems for missiles. Without military funding the computer revolution would not have taken place, and the superpowers' military capabilities owe as much to mathematics for their computer networks and control and communications systems as they owe to the nuclear physicists. It is time mathematicians felt the weight of this responsibility.

The received view of mathematics is that it is neutral with regard to both culture and gender. However, the divorce of mathematics from its social context, which this neutrality implies, leads to monoethnic, sexist, and possibly even racist, mathematics. The mathematics taught in schools and universities is too often presented as the product of white western men.

The third of the ways in which mathematics can be said to be isolationist is through its dehumanization. Mathematics can be said to be dehumanized in that the received views of the nature of mathematics, namely the classical philosophies of mathematics ignore the role of humanity in creating mathematics. The traditional philosophical questions addressed are: What are the objects of mathematics? What is the basis of mathematical truth? These very questions presuppose an objectified and inhuman mathematics, mathematics purely as product. Of the traditional philosophies only intuitionism admits the process aspect of mathematics, and only in the stylized form of intuitionistic proofs and computations.

The fourth and final area in which mathematics may be said to be isolated or separatist is in its use to dehumanize social issues. One of the characteristics of mathematics is the intrinsic role of abstraction. Mathematical abstraction provides a metaphor for the stripping away of the softer human and moral aspects of social issues, which occurs so often in modern society. A training in mathematical and abstract thought allows social decision-makers to consider and adjudicate social issues without feeling the human consequences of their decisions.

Beyond the metaphor, mathematics provides the conceptual tools for the dehumanization of social issues. This usually takes the form of a focus on quantity instead of quality. In human terms we have the focus on IQ instead of intelligence and the use of numerical scores to represent academic achievement. In the realm of economics we have the focus on rates of inflation, M3, percentage wage increases, profitability and cost accounting, productivity and the colossal financial calculations of the IMF. The military have coined such terms as 'megaton' and 'megadeath'.

This list only samples some of the numerical concepts used to strip social issues of their human and moral dimensions. These concepts are used to make decisions that materially affect the lives of millions of people without the impingement of the human or moral dimension. Consider only the concept of 'megadeath'. The concept of a million — a wholly abstract number — interposes itself between our comprehension, and the horror of the meaning of the term.

Gilligan has distinguished between 'separatist' and 'connectivist' modes of moral reasoning.[2] The separatist approach relies on the abstracting metaphor and concepts discussed above. In contrast the 'connectivist' approach explores the full social context surrounding a moral situation. What I am proposing is that a more 'connectivist' approach to the social significance of mathematics be adopted, or at least considered. Can we hide behind the argument that we only provide the metaphor and concepts for the dehumanization of social issues and that therefore our hands are clean? If providing such tools leaves the tool-maker free from responsibility, is not the same true for weapons manufacturers?

Mathematics has for too long isolated itself in the classroom, in the universities and away from the world. It is time we recognized that we are a small part of an enterprise much greater than ourselves, namely humanity.

This chapter constitutes a contribution to a discussion of the social and political values underlying mathematics and mathematics education which are so often ignored or denied. I have offered some controversial assertions in the hope of provoking a wider discussion of these vital but neglected issues.

Notes

1 DEPARTMENT OF EDUCATION AND SCIENCE (1981) *Aspects of Secondary Education in England: Supplementary Information on Mathematics*, London, HMSO.
2 C. GILLIGAN (1982) *In a Different Voice*, Cambridge, Mass., Harvand University Press.

18 *The Politics of Numeracy*

Jeff Evans

What Is Numeracy, and Why Is It Important?

'Numeracy' is a term which has taken its place, if not in the public consciousness, then at least in the language of circles where education is discussed, in the 1980s in Britain — following the publication of the Cockcroft Report, *Mathematics Counts* (1982). What meaning does this term have?

The term 'numeracy' was coined 'to represent the mirror image of literacy', by the Crowther Committee (1959, pp. 269–70). The Cockcroft Report discusses a range of definitions: from Crowther's broad conception — including familiarity with the scientific method, thinking quantitatively, avoiding statistical fallacies — to narrower ones, e.g., the ability 'to perform basic arithmetic operations' (*Collins Concise Dictionary* — and for many in the public at large?)

Taking an intermediate position, Cockcroft uses the word 'numerate' to mean the possession of two attributes:

1 an 'at-homeness' with numbers, and an ability to make use of mathematical skills which enables an individual to cope with the practical mathematical demands of his everyday life;
2 an appreciation and understanding of information which is presented in mathematical terms, for instance in graphs, charts or tables or by reference to percentage increase or decrease.

That is, they are concerned with the wider aspects of numeracy, and 'not ... merely ... the skills of computation' (Cockcroft, paragraph 39, p. 11).

There are several noteworthy aspects of this definition. First, both attitudes — an 'at-homeness' — and skills are considered important: *confidence* counts, as well as competence. Second, the touchstone for which skills are important is *practical*, and the relevant context is provided by the demands of the person's everyday life. Third, attention is directed to the appreciation of numerical information as well as the use of techniques, and this appreciation is implicitly *critical* or sceptical.

Several examples may illustrate the noteworthy features of this definition of numeracy: confidence, practicality, and its critical potential.

First is a quote from a series of reflections on schooldays by Margaret Drabble, a well known British writer:

> I dropped mathematics at 12, through some freak in the syllabus . . . I cannot deny that I dropped maths with a sigh of relief, for I had always loathed it, always felt uncomprehending even while getting tolerable marks, didn't like subjects I wasn't good at, and had no notion of this subject's appeal or significance.
>
> The reason, I imagine, was that, like most girls I had been badly taught from the beginning: I am not really as innumerate as I pretend, and suspect there is little wrong with the basic equipment but I shall never know.
>
> . . . And that effectively, though I did not appreciate it at the time, closed most careers and half of culture to me forever. (*The Guardian*, 5 August 1975, p. 16).

Second is an example where an apparent lack of numeracy had substantial practical consequences. On Saturday 23 July 1983.

> A simple metric mixup . . . nearly cost the lives of 61 passengers and eight crew members aboard an Air Canada Boeing 767 . . . The airline admitted that the fuel for Flight 143 from Montreal to Edmonton was calculated in pounds instead of kilograms (resulting in less than half the fuel necessary for the trip) . . . . The incident may re-kindle opposition to metrication. (*Toronto Star*, 30 July 1983, pp. 1 and 4).

Finally, this example concerns the critical appreciation of information. In both the 1983 and 1987 British elections one of the issues has been the reductions in public expenditure — whether there really have been 'cuts' in particular areas, and, if so, whether there have been reductions in actual services — rather than just the 'elimination of bureaucratic waste' (as the supporters of the cuts often put it).

In Britain people are particularly concerned that the National Health Service should be preserved. In the run-up to the 1983 election the Secretary of State for Health claimed that the NHS would have grown between 1979 and 1984 by 7.5 per cent. As a leading newspaper commented, 'The Government claims that it has increased spending on the National Health Service. Everyone else seems to believe otherwise' (*The Guardian*, 14 March 1983). In 1987 a major issue has been whether or not the government has maintained the level of funding for state education. In both cases critical analysis would require the confidence to seek out the relevant information, and to challenge government assumptions on issues like the following:

1 what level of inflation is appropriate to use for changing costs of health or education over time;
2 what are the best indicators of relevant provisions, e.g., spending per student, pupil-teacher ratio, the 'capitation allowance', etc.;
3 how 'needs' are changing, e.g., because of falling rolls, or the ageing population; and
4 which disaggregations are relevant, e.g., current versus capital spending, secondary versus primary versus nursery levels?[1]

The mention of assumptions about the rate of inflation is significant. The government's claims to have reduced the rate of inflation, discussion of what the levels of unemployment actually are, and debates on the need for Britain to have nuclear weapons, in the context of the actual balances of forces in Europe, have been major political issues for some time — and all are based to some extent at least on numerical information which must be interpreted critically.

The Politics of Numeracy

A number of people have written about the 'politics' of numeracy. Sometimes the concern is with the impact of the social and political values implicit in the traditional curricula and pedagogy (e.g., Ernest, 1986). Sometimes the focus is on inequalities in the distribution of numeracy, or the lack of it, 'non-numeracy' — and the effects of these inequalities.

In line with this second focus, this chapter discusses questions such as: Which social classes, gender groups, races, etc. benefit in terms of 'getting more than their share' of numeracy, and which lose in terms of being 'deprived' of numeracy? What advantages flow from being numerate, and what disadvantages from lacking numeracy? What are the ideological, as well as material, consequences of whatever inequalities there may be in the distribution of numeracy?

Non-Numeracy among Adults

There is a great deal of information available on the measured skills of school children of various ages, but relatively little on adults' levels of skills, with the exception of two surveys to be discussed in more detail below.

Figure 1. Questions 4–7 in the National Gallup Survey of Adults

4. Which is bigger, three hundred thousand or a quarter of a million? (Read out and show CARD 4)

ANSWER (write in): ....

Method: 1 Oral
2 With writing
3 Calculator used

Response: 4 Confident
5 Unconfident

6 Immediate
7 'Pause for thought'

BREAK

5. If you buy five Xmas cards for 65p, how much is each card costing you? (Read out and show CARD 5)

ANSWER (write in): ....

Method: 1 Oral
2 With writing
3 Calculator used

Response: 4 Confident
5 Unconfident

6 Immediate
7 'Pause for thought'

7. Suppose that the rate of inflation had dropped from 20% to 15%, which one of these results would you have expected:

(a) Prices would have gone down, or

(b) Prices would have stayed the same, or

(c) Prices would still be rising but not as fast as before, or

(d) Prices ought to have gone down but didn't

6. Here is a railway timetable. I live in Leicester and have arranged to meet a friend at the station in London at 4 o'clock in the afternoon. Assuming the trains run on time which is the *latest* train I can get from Leicester to arrive in time for the meeting?

Mondays to Fridays

Leicester dep.	*London* arr.
01.36	03.52
02.20	05.22
05.00	07.34
06.17	08.18
06.52	08.47
07.17	09.02
07.33	09.12
08.07	09.45
08.23	09.50
08.34	10.11
08.55	10.36
09.11	10.45
09.33	11.36
10.22	12.06
10.40	12.50
11.27	13.08
11.42	13.40
12.27	14.08
12.48	14.59
13.25	15.02
13.44	15.42
14.27	16.10
14.42	16.52
15.31	17.13
15.44	17.42
16.27	18.08
17.13	18.51
17.28	19.10
17.53	19.55
18.27	20.05
19.30	21.03
19.41	21.42
20.30	22.04
21.24	23.31

The Survey of Adults for Cockcroft

The Gallup survey which formed part of the evidence submitted by the Advisory Council for Adult and Continuing Education (ACACE) recruited a sample of almost 2900 adults in February 1981.[2] All eleven of the questions were meant to test everyday or 'practical' maths; Questions 4 to 7 are given as illustration in Figure 1. Overall six of the questions had to do with shopping or eating out, e.g. Question 5. Perhaps three of the questions were rather more 'formal': Questions 4 and 9a and 9b which required the reading of a graph about temperature changes. On the other hand, Questions 6 and 7 were about reading a railway timetable and

Table 1. Results of Gallup National Survey on Numeracy Skills

CORRECT ANSWERS (PERCENTAGES) ANALYSED BY SEX, AGE AND SOCIAL CLASS

		SEX		AGE					CLASS			
	Total	Men	Women	16-24	25-34	35-44	45-64	65 +	AB	C1	C2	DE
Base	2,890	1,385	1,505	529	581	493	812	476	463	636	943	848
Question 4	77	87	67	73	83	79	78	70	89	82	78	65
Question 5	68	72	65	66	75	70	67	64	84	77	65	58
Question 6	55	61	50	59	68	63	52	34	79	67	53	37
Question 7	40	45	36	48	51	44	34	28	60	49	38	29

CORRECT ANSWERS (PERCENTAGES) ANALYSED BY TERMINAL EDUCATION AGE

		TERMINAL EDUCATION AGE				
	Total	14 or under	15	16-18	19-20	21 or over
Base	2,890	813	750	1,056	73	199
Question 4	77	70	74	79	88	93
Question 5	68	60	65	72	92	90
Question 6	55	37	53	64	82	85
Question 7	40	27	35	47	69	72

understanding 'inflation', respectively. The results analyzed by sex, age and social class are given in Table 1.

We can summarize the results as follows:

in questions on simple operations percentages were answered correctly between 64 and 88 per cent of the time, but Questions 6 and 7 were answered correctly less than 60 per cent of the time;

men did (2–20 per cent) better than women, and the difference is largest on Questions 4, 6 and 3 (to do with calculating a 10 per cent tip on a restaurant bill);[3]

the young, especially the 25–34 age group, generally do better, and the over 65s least well, though the difference is less on 'shopping' questions like Question 5;

social classes AB (professional and intermediate occupations) do best, and classes DE (semi-skilled and unskilled) do least well; the higher the terminal education age (another indicator for social class), the better the level of performance.

This survey provides further information on the three aspects of numeracy discussed above. Thus, in addition to the responses themselves, interviewers recorded whether each response was made in a 'confident' or 'unconfident' way, and whether 'immediately' or after a pause for thought. In general, the higher the proportion of correct answers in a group, the more confidently and immediately the answer was given.

Though posed in a formal interview situation, all of these questions involved reasoning which might be required in an everyday situation rather than merely performing a purely abstract sum.[4] None of the questions required skill in critically assessing numerical information before doing calculations with it. However, Question 7 addresses an issue of considerable political and public policy importance: the meaning of the notion of 'inflation'. It was by far the least well answered question. Any of the wrong answers on this question indicate that the respondent will be substantially confused over an issue which has been widely discussed over the last twenty years, and which the government has specified as a top priority. Of the 54 per cent of people answering this question incorrectly, about one-third (or 17 per cent of the whole sample) gave response (d): 'Prices ought to have gone down, but didn't.' This particular incorrect answer is clearly based not only on a confusion between the *level* of prices and the *rate of price rises* (as in response (a)), but also on despair that their expectations are not satisfied. This might lead the respondent to be cynical about statistics and politicians and perhaps the media in a way that is, in this particular case, apparently not deserved.

The Survey of 23-Year-Olds

Another study used data from the fourth 'follow-up' of the National Child Development Study, which interviewed some 12,500 British 23-year-olds (ALBSU, 1983). It found that 5 per cent reported problems with numeracy ('number work' or 'basic maths'), as compared with 10 per cent for literacy (reading, writing and spelling), and 2 per cent for both. Of the 5 per cent reporting problems with numeracy, over a quarter reported difficulties in everyday life arising from these problems.

In seeking to reconcile the results from the two surveys, we note the following. The ACACE's results are based on answers to eleven questions given by an interviewer, whereas the ALBSU/NCDS study is based on the individual's own assessment of his or her skills. The 95 per cent reporting no problems in the latter should be compared with a 73 per cent average of correct answers for the 16–24 age group on the former: the fifth percentile for this age group is only three questions correct. It appears that ALBSU's respondents' self-ratings of their numerical skills may have been over optimistic. Willis (1983) echoes this scepticism: 'Can this really be true? Only 1 in 20?'

Some further evidence is available. The ALBSU was quoted during National Numeracy Week in 1983 (*TES*, 9 September) as believing that one in four adults are not able to calculate change from £5 for one item. More recently one in four of a University of Lancaster sample of 500 teenagers and 500 adults was reported as unable to work out that thirteen £5 notes made up £65; and the Manpower Services Commission estimates that 20 per cent of the long-term unemployed have some kind of literacy and numeracy problems (*The Guardian*, 14 May 1987).

These two surveys are the only known large-scale ones which allow us to assess adults' skills in numeracy in the UK; further, they allow us to estimate inequalities in this area related to social class and gender.

Interviews with Adults

Brigid Sewell was commissioned by ACACE and the Cockcroft Committee 'to provide evidence and information on the mathematical needs of adults in daily life' (1981, p. 1). Sewell decided to interview members of her sample twice: the first time to 'ease tension', to discuss selected situations in which maths might be used (numerical 'needs'), and the respondent's attitude to mathematics; and the second time to discuss in detail strategies for solving problems chosen for their common relevance, and the solutions.

Besides indicating how widespread were various needs for numeracy, the first interview produced some indicators of attitudes in two senses. First, the refusal rate for that interview was about 50 per cent (p. 11), and Sewell attributed this to people's perceptions of maths as a 'daunting subject' (p. 11). Second, in answer to Question 22: 'Do you enjoy working with numbers?', half the sample said 'yes' and the other said 'no' 'with varying degrees of antipathy'; in answer to Question 23: 'How well would you say you can manage in everyday situations when numbers are involved?' 76 per cent answered 'very well' or 'all right', 18 per cent 'mostly' and 5 per cent 'with difficulty'. The unsolicited remarks about the experience of maths were much more negative (p. 16).

For the second interview a number of respondents, considered to have interesting patterns of experiences with, and perceptions of, maths, were selected from three bands of competency, depending on their facility with percentages at the first interview. The level of correct responses was, if anything, lower than on the Gallup survey (ACACE, 1982, p. 42). For example, the answers to the question on inflation, similar to Question 7 on the Gallup Survey, were: 32 per cent correct; 44 per cent wrong (e.g., falling inflation means prices should fall, but they don't: 'it's all a big con trick!'); 14 per cent don't know; 10 per cent ambiguous or incomprehensible.

Any apparent differences would need to be interpreted in the light of differences in the way the data were produced, such as:

1 the interview questions were more 'practical', using, for example, wage slips, maps and electricity bills;
2 sampling methods, non-response rates, and further selection for the second interview (ACACE, 1982, p. 40);[5] and
3 differences in the survey and interview situations.

What Are the Consequences of Low Levels of Numeracy among Adults?

My purposes here are to sketch the sorts of consequences which flow from lacking numeracy. Here, I shall organize my discussion on the basis of distinguishing individual and societal levels, and also 'material' and the ideological spheres.

Consequences at the Material Level for an Individual

These have to do with restrictions on one's freedom of access to further education and training, restriction on access to jobs (and the related rewards of income, companionship, sociability, satisfaction, etc.), and also self-restriction in the form of subject choice (and job choice). They also have to do with the ability to perform in — and to enjoy — one's job and everyday life. Thus non-numerate adults will tend to avoid training or courses which seem to them to involve maths, even if they are not already barred by their lack of qualifications, or performance on a TOPS test, and they may tend to skip over graphs in a newspaper, or tables of figures in a budget.

In considering access to education and training, we need first to consider the formal measures which are used by institutions to offer or refuse such access. The pre-eminent qualifications which are used (validly or otherwise) as indication for numeracy in Britain are GCE exam results — O-levels, and in some cases A-levels.

Concerning access to higher education, I considered social science degree courses in universities and polytechnics, and the BEd. *Which University*? shows that the overwhelming number of social science degree courses require O-level maths for entry. For BEd. degrees the DES has ruled that, from the 1981 intake, all students shall have O-level maths, except for mature students who sit a special entry test at the time of selection. In the case of access to jobs for school leavers, the *Careers Advisers' Handbook 1982–83* indicated over 200 jobs; of the sixty-five or so jobs mentioned as having a GCE qualification, fifty-four include maths qualifications, compared with forty including English.

This message is emphasized by Lucy Sells (1978) in her 1972 survey of educational and vocational opportunities for American women.

> Without four years of high school maths, students at Berkeley were ineligible for the calculus sequence, unlikely to attempt chemistry or physics, and inadequately prepared for intermediate statistics and economics. Since they could not take the entry level courses in these fields, 92 per cent of the females would be excluded from ten out of twelve colleges at Berkeley and twenty-two out of forty-four majors. (Tobias, 1978, p. 13)

Sells is cited as arguing that

> knowledge of algebra and geometry divides the unskilled and clerical jobs from the better-paying, upwardly mobile positions available to high school graduates. She estimates that mastery of high school algebra alone will enable a high school graduate to do so much better on a civil service or industrial exam. . . . Just one more year of high school maths could make the difference between a starting salary of $8000 and one of $11,000. (reported in Tobias, 1978, p. 26)

The generally accepted finding is that income depends on the *number of years* of schooling. What is being suggested here deepens this to say that *what* is studied — in particular, how much maths — matters too.

Sells' message is confirmed in the UK. First, let us consider male/female differences in subject choice. In examination entries in ILEA in 1976, for example, maths was third in number of O-level entries overall, behind English language and English literature, but among the major subjects maths had the fifth lowest percentage of female entrants. This may be both a consequence of developing gaps in numeracy between boys and girls, *and* a contributory influence on differing levels of numeracy between adult men and women.

Next we consider ethnic differences in the percentage gaining maths O-level among West Indians (5 per cent), Asians (20 per cent) and others (23 per cent); the corresponding figures for English were 9 per cent, 21 per cent and 34 per cent, respectively (Rampton Committee, 1981, p. 63).

So far a lack of numeracy has been captured almost entirely by a lack of school maths qualifications. The consequences of lack of numeracy for performance in jobs and in everyday life can also be documented graphically in other ways. For example, in the profiles of interviewees presented in Sewell (1981), Ian, a modern languages graduate in his 20s, was one of the intending OU students interviewed. He admitted avoiding numbers as much as possible: 'Numbers are anathema to me.' His reaction to the inflation question was that he hated it, did not know what inflation meant, did not know how it affected him; altogether he felt 'terribly out of it'. Sewell concluded: 'his lack of mathematical confidence ... has heavily influenced his choice of career and everything he does' (Sewell, 1981, pp. 48–9). This echoes Margaret Drabble's feelings about being excluded from one of the 'two cultures' (quoted earlier).

Material Consequences on the Societal Level

These include loss of production, in quantity or quality, waste of resources, production of inaccurate or useless information, and threats to life and limb, as in the airline mix-up described above. Examples can be found in the public policy-making area by the way clerical errors, once incorporated into the numbers, are allowed to pass without being detected. The following examples are cited in a discussion of the production of official statistics: an accidental omission of a zero by an Olivetti employee reporting the firm's exports generated a phoney balance of payments crisis; a clerk's copying two lines of figures onto a coding sheet in the wrong order caused the trade figures to go haywire over many months; the accidental counting of the same set of movements twice led to a major error in Home Office migration figures. The authors conclude: 'Serious errors would certainly occur less often if staff had the ability to recognize figures as implausible and the initiative then to get them sorted out' (Government Statisticians' Collective, 1979, p. 144). Doubtless readers can recall their own stories of this type.

Nor are these societal consequences likely to be confined to the public sector: low numeracy may well be one aspect of the problem of the alleged shortcomings of British management. It is argued by Lynn Osen, a mathematician, that business today needs people 'who can understand a simple formula, read a graph and interpret a statement about probability' (Osen, 1971; quoted in Tobias, 1978, p. 27). An illustration is given by Chris, in his 30s, who is managing director of his own building firm; he manages at work by 'bluff[ing] my way', and at home through his wife dealing with the domestic bills. At the time of his interview he was attending adult literacy and arithmetic classes as he wanted to be able to understand building plans and quantity surveyor's estimates (Sewell, 1981, p. 42).

Ideological Consequences at the Individual Level

These consequences of a low level of numeracy include not only a lack of competence among adults, but a low level of confidence in their constructive skills and critical insights. This leads them to be dependent on the views of the 'expert' or 'professional' for their opinions, and susceptible to the mystique of mathematics. This mystique derives no doubt from the conciseness of mathematics, its apparent precision, its obvious abstraction and strangeness (e.g., the use of many Greek symbols) and its association with modern science; and it tends to lead to the following ideas:

1 arguments involving numbers are (or tend to be) more 'rigorous' than those without;
2 this rigour means that in such arguments there is less room for debate (after all, there was only one right answer in maths at school — and usually only one way apparent to get it);
3 people who use numbers are (or tend to be) more rigorous, and hence trustworthy, than others;
4 such people have a 'mathematical mind'; if you don't there is no hope for you with numbers;
5 because maths is cumulative, if you fall behind, you'll never get a second chance to learn it.[6]

At the individual level the result of this diffidence and dependency is often that people lurch between two traps: uncritical acceptance of claims made, and an equally uncritical rejection, based not on a consideration of the evidence, but on prejudice or the unexamined authority or 'experts'. Sometimes the failure to consider the evidence comes from *reluctance to seek it*, sometimes from mistakenly *mistrusting* evidence which is available, sometimes from *misinterpreting it*.

An example where *evidence was mistrusted* by an individual comes from a prime-time television discussion about access to higher education, 'Inquiry: The Race for a Place' (BBC 2, Friday 4 November 1983, 7.30–9.00 pm). In response to presenter Ludovic Kennedy's claim that the UK was well behind its European partners in providing access to higher education, Sir Keith Joseph, the Secretary of State for Education, argued that, because of our shorter degree courses, differences in definition of courses, etc., we were not behind other European countries (though we were behind the US). The response of Ludovic Kennedy — not generally known as an unthoughtful or uncritical person — was to fall towards the second trap specified at the beginning of the previous paragraph (uncritical rejection): 'Well, Sir Keith, we all know you can prove anything with statistics. . . .'

Ideological Consequences at the Societal Level

When evidence is not sought, or when it is mistakenly mistrusted or misinterpreted *at the societal level*, we may sometimes speak of *myths* — ideas which are partially or largely false but which influence the beliefs and the actions of large segments of society, and affect the practices of society's institutions.

Some myths seem to have grown up fairly 'naturally', for example, myths around the role of women in areas such as employment, such as the following:

1 A woman's marital status is a crucial determinant of whether or not she works;
2 Most married women don't need to work;
3 Women leave work to have babies and don't come back;
4 In times of high unemployment women who work are taking jobs away from men;
5 In the current crisis women are becoming unemployed at a greater rate than men.

Ways of critically addressing these myths using evidence from official statistics such as the 'General Household Survey' are discussed in Lievesley *et al.* (1983). Despite the fact that the statistics on which these critical scrutinies are based are available in many local authority or college libraries, and are published by at least some national newspapers, the level of discussion of them in the media is rather low, and most members of the public are probably not aware of their existence, let alone of their content. Thus, there is great *reluctance to seek* evidence relevant to myths as important as 1–5 above.

Other myths seem to have to be fostered rather more actively. When people are convinced, through the use of 'scientific' or 'mathematical' arguments, to accept myths that it is against their interests to believe, we may speak of 'mystification' (see Irvine *et al.*, 1979, Introduction).

An example of mystification, showing *misinterpretation of evidence*, comes from the relationship between class size and pupil attainment. In a number of educational studies done over the last twenty years or so there appears to be a 'positive correlation' between the two, i.e., as class size increases (across different classrooms), the average level of attainment also tends to increase; or else there is found no relationship at all, i.e., as class size increases, attainment appears to remain relatively unchanged. Thus, the statistics seem to challenge what teachers know by 'common sense'.

Some people have tended to interpret the 'findings' as saying that we could pack at least a few more children into a particular classroom without any appreciable negative effects. Teachers are equally clear that increasing class sizes will interfere with many important processes in the classroom. Yet they may not be confident enough to challenge the interpretation of the statistics which ignores that 'correlation is not causation' and which fails to investigate alternative explanations for the correlation observed.[7]

Why Do Adults Have These Problems with Numeracy?

When we come to explain these sorts of problems with numeracy, we find that much of the research in mathematics education understandably focuses on:

1 school factors, including teachers, which affect performance in *school maths* (e.g., Bell *et al.*, 1983), with less attention to extra-school factors and processes. The latter are clearly important in forming adults' numeracy, and include:
2 home, parents and siblings;
3 out-of-school activities and peers;
4 post-school education;
5 work and work-training schemes; and
6 everyday life and adult subcultures.

We also find, as with other problems in the social sciences and education, that we are faced with what appear to be two different kinds of explanation. We might explain how and why people act as they do by reference to 'socialization pressures' which have moulded them in certain ways; or we might see them as acting in accordance with their 'perceptions and purposes'.[8] I do not believe the distinction between these two approaches is as neat as is sometimes maintained. We might say that socialization pressures are what make children what they are, and that a person attains adulthood to the extent that they are able to transcend such pressures, and decide to act on the basis of their perceptions and purposes. But this would be too simple: it would ignore the already developing autonomy of children, and it would forget that adults too are constrained by social forces: 'Men make their own history — but they do not make it in conditions of their own choosing': thus Marx tried to reconcile determination and freedom in his explanations of social action. The conceptual map I have developed therefore includes both socialization pressures and perceptions and purposes; a sketch is given in Figure 2.

Thus family and school provide socialization pressures, and 'personal characteristics' may act as a distillation of these pressures in the life history of the individual; 'motivation', 'interest' and 'attitudes' are involved as perceptions and purposes (though attitudes are also sometimes considered as personal characteristics). Work and post-school education could be seen as *resources* which may be used by the emerging adult working towards desired outcomes. We may also need to consider as societal *constraints* factors such as the following: the professional interests of mathematicians, political interests of, for example, public servants, and the mystique of mathematics (discussed above). All of these features may help to explain the progress of adults towards outcomes such as:

competence and skill in using numbers practically and critically/sceptically;
confidence with numbers rather than anxiety;
perseverance with using numbers — especially with enrolling for maths courses — rather than avoidance;
enhanced choice of courses of study and occupations; and ultimately richer personal development and experiences.[9]

Figure 2. Why Problems with Numeracy?

SOCIALISATION PRESSURES

HOME
-PARENTS
*own ability
*expectations
-SIBLINGS
*competition
*comparisons

SCHOOL
-TEACHERS
*methods
*own anxiety
-ATMOSPHERE
*ability gps.
*questioning

ACCIDENTS OF BIOGRAPHY
*chg. of schl. *illness

EVERYDAY LIFE
*activities
e.g.budgeting
games

ADULT
SUBCULTURE
*discussion

"PERSONAL CHARACTERISTICS"

ATTITUDES
*anxiety

MOTIVATION
*interests

POSITIONS
*gender
*soc.class
*ethnic

PEER SUB-
CULTURE
*labelling
e.g."swot"

POST-SCHL.
EDUCATION
*"second
chances"

WORK
*skill need
*encouraged
by workmates

OUTCOMES

QUALIFICATION
CONFIDENCE
PARTICIPATION
COMPETENCE
ACCESS - F/HE
ACCESS - JOBS

RESOURCES/
CONSTRAINTS

POL.INTEREST
PROFESSIONAL
INTERESTS
*mystique

Conclusion: What Is to Be Done?

Our discussion of 'numeracy' and of its consequences shows that a lack of numeracy is a disadvantage for most adults; it can further be considered a facet of the oppression of women, the working class and ethnic groups. This suggests a number of changes that could be made in the ways that maths is taught and numbers are used in order to help improve adults' numeracy.

1. 'Second chance' courses for adults seeking to develop their numeracy need to take account of the oppressive features of early maths experience, starting with their setting and their style: thus the importance of the informality and accessibility of centres or programmes aiming to attract working-class participation, e.g., the walk-in numeracy centres in Hammersmith or Edinburgh (Jordinson, 1987), or the ITV programmes such as 'Counting On' (September–November 1983). Similar conditions need to be specified for courses aiming to attract (mainly) women (see Tobias, 1978).
2. The process might start with a diagnostic interview in which a 'counsellor' asks about past experiences with numbers, aiming to build a mathematics autobiography (Tobias, 1978, pp. 250 ff.) or consciousness raising (see Frankenstein, 1983, for a relevant discussion of Paulo Freire's ideas on pedagogy). The aim would be to help the participant to understand better his or her experiences, to find out what the learner *does* know rather than what (s)he does not, and to formulate goals for the future.
3. Curriculum development should be linked with a discussion of the practical needs of adults, and of the maths actually used by adults in everyday life (Sewell, 1981) and at work (Cockcroft, 1982, Ch. 3).[10] As may be inferred from the discussion here, I consider statistics to be a promising context in which to involve many adults in numeracy (see also Evans, 1986).
4. In our pedagogy we can challenge the following assumptions and practices that have made maths formidably off-putting:
   - (i) it is cumulative and strictly ordered: there are *many* interesting places to start (see Tobias, 1978, Chs 6 and 7);
   - (ii) there is always one correct route to one correct answer and no room for discussion: we can teach with seminars where numerical issues are debated (see, e.g. Frankenstein, 1983), and give assignments that are really essays reporting the sometimes tortuous path of an 'investigation';

(iii) maths involves struggling alone, often under the eye of the teacher: we can use group work (Frankenstein, 1983), and also an element of self-pacing exercises (using microcomputer resources);

(iv) maths teachers are often oppressive as human beings: we can see confidence-building as one of our major aims — both to contribute to personal development of our students and to help them sharpen their critical potential.

5 In assessment we can challenge the impression that maths involves doing timed tests made up of abstract sums. Besides emphasizing practicality we can use project assessment and investigations.

6 We can try to insist on — and contribute to — clarity of presentation of numerical material in the media. The experiences of the Open University broadcasts (e.g., those for 'MDST 242: Statistics in Society') — plus those in popular broadcasting, e.g., Professor Bob MacKenzie's 'swingometer' to dramatize the effects of changes in percentage of the electoral vote — show that where ingenuity is applied, clarification of people's ideas may well result.

7 We can help adult students, and our fellow citizens, to avoid both 'Scylla', uncritical acceptance of numerical arguments, and 'Charybdis', uncritical rejection, in several ways:

by using our skills as maths educators to show that the arguments are not *inaccessibly* technical;

by reassuring them that the arguments are not *wholly* technical (as, for example, when 'needs' are part of the argument, as in myth (2) about working women above); and

by fostering people's confidence to make critical assessments.

For example, 'analyses of government spending can be carried out by anyone with access to an appropriate library; expert knowledge is not needed' (Radical Statistics Education Group, 1987).

Acknowledgment

I should like to thank John Bibby, Len Doyal, Paul Ernest, Harvey Goldstein, Eva Goldsworthy, Ken Menzies, Jos ter Pelle, and Valerie Walkerdine for helpful comments on an earlier draft of this chapter.

Notes

1 For further discussion of how to scrutinize critically government spending in education and in health, see RADICAL STATISTICS EDUCATION GROUP (1987) and RADICAL STATISTICS HEALTH GROUP (1987), respectively.

2 Sampling took place at 200 sampling points in ten regions in England, Wales and Scotland. The interviewers were given quotas for sex by age, social class and employment of respondents.
3 These results are subject to sampling variation (see ACACE, 1982, p. 9); thus any difference between the male and female subgroups of 4 per cent or less would not be impressive (since it could be expected to occur, due only to chance, nineteen times out of twenty).
4 This distinction between formal and everyday contexts is parallel to that made between 'folk maths' and 'school maths' in MAIER (1980), which gives approximation as an example of a skill used much more in everyday contexts than in formal ones.
5 The sampling method for the first interview might be called 'multiple snowball recruitment' with the snowballs starting from the inquiry officer's friends, colleagues, adult numeracy classes, WEA class, and an Open University introductory course (for the Arts Foundation Course). See also Note 2.
6 These arguments are discussed in TOBIAS (1978, *passim*).
7 For further discussion of these issues see SIMPSON (1980) and RADICAL STATISTICS EDUCATION GROUP (1982).
8 See PRING (1980) for a discussion of these issues in relation to the Rutter research.
9 For further development of this conceptual map, and its use in a study of 'maths anxiety' among adult students, see EVANS (forthcoming).
10 See RILEY (1983) for cautions against uncritical adoption of the goal of 'functional numeracy'; see also HARRIS (1982) for scepticism that employers are clearly aware of the needs of the jobs in their establishments.

References

ADULT LITERACY AND BASIC SKILLS UNIT (ALBSU) (1983) *Literacy and Numeracy: Evidence from the National Child Development Study*, London, ALBSU.

ADVISORY COUNCIL FOR ADULT AND CONTINUING EDUCATION (1982) *Adults' Mathematics Ability and Performance*, Leicester, ACACE.

BELL, A.W., COSTELLO, J. and KUCHEMANN, D. (1983) *A Review of Research in Mathematical Education, Part A: Research on Teaching and Learning*, Windsor, NFER-Nelson.

BUXTON, LAURIE (1981) *Do You Panic About Maths? Coping with Maths Anxiety*, London, Heinemann.

Careers Advisers' Handbook (1983) London, New Opportunity Press.

COCKCROFT, W.H. (1982) *Mathematics Counts*, London, HMSO.

CROWTHER COMMITTEE (1959) *15 to 18*, London, HMSO.

ERNEST, P. (1986) 'Social and Political Values', *Mathematics Teaching*, 116, September, pp. 16–18 (reprinted as Chapter 17 in this book).

EVANS, J. (1986) 'Statistics and Numeracy for Adults: The Case for the Barefoot Statistician', Contributed paper for the International Conference on Teaching Statistics, Victoria, Canada, August.

EVANS, J. (forthcoming) 'Only Girls Are Anxious: Femininity, Masculinity, and Maths Anxiety', in V. WALKERDINE (Ed.), *Girls and Mathematics: New Thoughts on an Old Question*, London, Methuen.

FRANKENSTEIN, M. (1983) 'Critical Mathematics Education: An application of Paulo Freire's Epistemology', *Journal of Education*, 165, 4, pp. 315–39.

GOVERNMENT STATISTICIANS' COLLECTIVE (1979) 'How Official Statistics Are Produced: Views from the Inside', In Irvine *et al.* (1979).

HARRIS, M. (1982) 'Turn Upsidedown and Multiply', *Struggle: Mathematics for Low Attainers*, 9. ILEA.

IRVINE, J., MILES, I. and EVANS, J. (Eds) (1979) *Demystifying Social Statistics*, London, Pluto Press.

JORDINSON, R. (1987) 'The Edinburgh Walk-in Numeracy Centre (EWINC)', Discussion Paper 16, Association for Recurrent Education, Nottingham.

LIEVESLEY, D. *et al.* (1983), 'Myths about Women: How Statistics Can Help', *Radical Statistics Newsletter*, 27, May, pp. 5–7.

MAIER, E. (1980) 'Folk Mathematics', *Mathematics Teaching*, 93, December.

OSEN, L. (1971) *The Feminine Math-tique*, Pittsburgh, KNOW.

PRING, R. (1980) 'Cause and Effect', in *The Rutter Research, Perspectives 1*, University of Exeter, School of Education.

RADICAL STATISTICS EDUCATION GROUP (1982) *Reading between the Numbers: A Critical Guide to Educational Research*, BSSRS (25 Horsell Rd, London N5).

RADICAL STATISTICS EDUCATION GROUP (1987) *Figuring out Education Spending: Trends 1978–85 and Their Meaining*, BSSRS.

RADICAL STATISTICS HEALTH GROUP (1987) *Facing the Figures: What Is Really Happening to the National Health Service*, BSSRS.

RAMPTON COMMITTEE (Committee of Enquiry into the Education of Children from Ethnic Minority Groups) (1981) *West Indian Children in Our Schools: The Interim Report*, London, HMSO.

RILEY, T. (1983) 'Functional Numeracy', *Numeracy, Viewpoints 1*, London, ALBSU, pp. 2–4.

SELLS, L.W. (1978) 'Mathematics: A Critical Filter', *The Science Teacher*, February, pp. 28–9.

SEWELL, B. (1981) *Use of Mathematics by Adults in Everyday Life*, Leicester, ACACE.

SIMPSON, S. (1980) 'Wrong Questions, Wrong Tools', *The Times Educational Supplement*, 18 April, p. 13.

TOBIAS, S. (1978) *Overcoming Maths Anxiety*, Boston, Mass., Houghton Mifflin.

Which University? 1976, London, Haymarket Press.

WILLIS, J. (1983) 'Who Are These People with Numeracy Problems?', *Numeracy, Viewpoints 1*, London, ALBSU, pp. 19–21.

19 *Mathephobia*

Jenny Maxwell

'Mathephobia is irrational and impeditive dread of mathematics' (Lazarus, in Resek and Rupley, 1980). It was through Myrtle that I first recognized and became interested in the condition. I knew that many people disliked mathematics or found it difficult, but until I met Myrtle I was unaware of the terror and panic it can arouse. Myrtle was a student for five years in the adult class which I taught. She was an articulate lady whose achievement in other subjects was clearly much greater than in mathematics. This discrepancy was obvious to her, and for a year or more the overwhelming necessity of hiding her difficulties from the rest of the class impeded any learning. With adults it is relatively easy to build a relaxed and communicative atmosphere, and we all helped Myrtle by sharing tales of beatings for not knowing tables and of unfulfilled parental expectations. After five years she was still apologetic for her mistakes, but she did recognize that they are common and can be fruitful. Above all she enjoyed mathematics.

It is possible to be a competent mathematician who dislikes the subject. The mathephobic's stomach churning fear and panic produces total inability to do mathematics. Inability is relative to expectations so that for many in the lower ability ranges the incomprehensible fog of mathematics merely confirms, and is confirmed by, their other experiences. It does not produce mathephobia. Laurie Buxton (1981) chose his research sample to be people who 'panic[ed] about maths' in spite of, or perhaps because of, achieving success in other fields. His most vivid example is of a headmistress who likened the panic when her husband was late home from a business trip to Israel to that experienced as a child when she was unable to remember 7×7. Women arts graduates are particularly prone to mathephobia, though both Laurie Buxton (1981) and Bridgid Sewell (1981) found that men were more reluctant to admit their fears, so there are probably more male mathephobics than would appear from their statistics.

A certain amount of anxiety is necessary for the educational task of learning mathematics. What is needed is previous success combined with just enough anxiety and pride to drive towards activities which are thought to be within reach. Aiken (1970) gave students a questionnaire asking them to write true or false by each of three statements:

> I am often nervous when I have to do arithmetic
>
> Many times when I see a maths problem I just freeze up
>
> I was never as good in maths as in other subjects.

His results accord with those of Janet Morris (1981), who speaks of students displaying 'panic, tension, helplessness, fear, distress, shame, sweating palms, clenched fist, queasy stomach, dry mouth, cold sweat.' Resek and Rupely (1980) speak of 'knots in the stomach and throbbing in the head at the mention of fractions or variables.'

About half of Brigid Sewell's (1981) intended sample refused to be interviewed. She attributed this to 'the painful associations which they feared [she] might uncover.' Of those who did agree to be interviewed the 'perception that mathematics is daunting pervaded a great deal of the sample selections.' During the interviews people chain-smoked, and there was much nervous laughter.

One of Laurie Buxton's (1981) subjects distinguishes between the emotional and physical reactions to mathematics. 'Panic in the mind and emotional feeling in the stomach are distinct. The mind itself is thrown into a confusion and cannot make the necessary connections .... Perhaps ... I haven't asked for the right [message]. I haven't made the right demand of my brain and the brain knows it and ... rebels.' To solve a problem the brain makes a plan and follows it through. The process can go wrong at each of its three stages. The expectation of failure can be so great that the brain cannot even begin to plan, it can try to plan but fail, or it can make a plan which it cannot execute. The second and third of these cause frustration. To the first, which causes panic, mathematical problems are particularly susceptible. An English undergraduate is unlikely to interpret the problem on a university mathematics paper. The mathematics undergraduate would have a better chance of making sense of an English question and, given library facilities, would have at least a rough idea of how to approach its solution. This peculiar characteristic of mathematics is recognized universally. Teachers whom I interviewed (Maxwell, 1984) agreed almost unanimously that 'in the maths department [they had] a little extra aura as far as the other staff [were] concerned.' While most thought such reverence misplaced and undeserved, they nevertheless thought it was a view generally held by the public.

Mathematics lends itself particularly to an authoritarian teaching approach which fosters this mystique, creating fear and panic. Many children are encouraged to accept the teacher's word without question, even when she appears to be behaving oddly. A language graduate told Bridgid Sewell (1981) that on starting algebra (for many children the end of meaningful mathematics) he failed to understand why a mathematics teacher should be writing letters. He was so accustomed to mathematics being incomprehensible and to the teacher's infallibility that he attributed all blame to himself, a view likely to be endorsed by the teacher.

It is hard for a teacher to empathize with a learner, but important, for a small piece of mathematical misunderstanding can support an edifice of nonsense.

> Take the number 10. We are so used to it that we cannot imagine [being] told that when you put (1 and 0) together, it stands for something much bigger than either of them. We should acknowledge the . . . nuttiness of this . . . so [the children] will not feel on the outside of a baffling mystery. Otherwise their first encounter with 10 may give children a shock from which they never fully recover and which freezes up their minds every time they think about it. (Holt, 1969)

Students are helped by teachers who can admit their own mistakes. A teacher was asking fourth year juniors to make given amounts of money using a minimum number of coins. As an example she gave 30p = 3 × 10p. The most disruptive child remembered what the teacher had forgotten — the new 20p piece. Her praise and the admission of her mistake had many positive consequences. The class relaxed, the boy was pleased, they were reminded that mathematical success is attainable by ordinary people, that mistakes are not fatal but inevitable and that this teacher welcomed questions and discussion.

Mathematics is seen by some as an infallible truth with absolute standards of right and wrong, words conveying moral values which extend beyond the classroom. It is more likely than any other subject to be taught in a way which 'hardly permit[s] a doubt or a suggestion from [a] student' (Joffe, 1981). It is over 100 years since Isaac Todhunter wrote, 'If [a schoolboy] does not believe the statements of his tutor . . . his suspicion is irrational and manifests a want of the power of appreciating evidence, a want fatal to his success in [mathematics]' (Griffiths and Howson, 1974, p. 296). A century later W.H. Auden (1973) recalls being taught: 'Minus times minus equals plus/The reason for this we need not discuss.' This equation, he says, was 'traditionally imparted by the rule of authority.' It should be possible for students to question and discuss such knowledge until it seems reasonable. One of Laurie Buxton's (1981) subjects says, 'You are forced to accept something when you don't want to, the whole of you revolts against it.'

Written constructive comments instead of a tick or a cross could remove the judgmental element from mathematics. So could the recognition that a variety of methods and even answers is possible. The answer to the problem, 'how could we best get to Birmingham city centre to go shopping from the university two and a half miles away?' could include consideration of speeds, parking facilities, costs of bicycle, car, bus, train or walking, producing many different possibilities. Discussion between tutor and students, including peer tutoring, is the basis of the 'Math without Fear' project in San Fransisco (Resek and Rupley, 1980).

John Holt (1969) believes that society conditions teachers into deliberately frightening children. Of an otherwise intelligent boy who is unable to learn simple number bonds he says:

> His memory does not hold what he learns, above all else because he won't trust it .... How can you trust any of your own thoughts when so many of them have proved to be wrong? .... We have made him afraid, consciously, deliberately, so that we might more easily control his behaviour and get him to do whatever we wanted him to do.

Whether or not you agree with Holt, it is certainly true that 'society' produces examination systems. These not only dominate curricula but introduce time constraints. Many people trace their unhappy memories of mathematics to their early years when they were required to give rapid mental answers. A friend remembers, like Myrtle, the fear and anticipation of physical pain causing total paralysis of her brain. After twenty-five years she has not recovered from her terror of mathematics. Some years ago at a day conference for adult numeracy tutors we were given a test of thirty mechanical questions including fractions, decimals and percentages, to be answered in ten minutes. It was of a type given to people to enrol on job training schemes. It needed little imagination to see how such a test could induce paralysing panic. My own anxiety was that I should not fulfil the expectations that I and others in the room had of me. The relief at avoiding the 'anti-goal' (Buxton, 1981) of failure was heightened by the knowledge that most other people had thirty right answers too.

For many children mathematics is a series of such anti-goals to be avoided. Emphasis on speed and accuracy creates the view of mathematics as an 'answer centred rather than problem centred subject' (Holt, 1969). Instead of learning mathematics children develop devices such as copying for producing right answers and 'defense mechanisms to protect themselves from defeat' (Holt, 1969). Two junior children in a special unit honestly thought that all that was required was a line of right answers and were surprised to be told that copying (as opposed to genuine helping, a most valuable way of learning for both helper and helped) was an inappropriate way to produce them.

Part of the teaching skill in avoiding fear is finding the right level of question. One of Laurie Buxton's (1981) group was succeeding until she was presented with the Wassan problem. Her increasing confidence was badly shaken by her failure to solve it. An adult student once came to my class wanting to multiply two digit numbers. He had no concept of tens and units, but when I tried to take him back to this he was deeply offended, his confidence shattered, and he never came back.

Mathephobics attach much blame to those teachers whom they have found to be impatient and unsympathetic, who shout or rely on fear or physical punishment to motivate their students. Children will accept that punishment is just for idleness or naughtiness but not for misunderstanding. It is more comfortable for a teacher to attribute a student's failure to inattentiveness than to an inadequate explanation.

Many teachers, often ill-trained and teaching mathematics unwillingly, see it as cold, impersonal and subhuman. It is a barrier to, rather than a means of, communication, which is not only about methods, hypotheses and answers but about feelings. Communication takes time – to play, relax, discuss, absorb, assimilate and understand. Lynn Joffe (1981) working with dyslexic children spends the first ten minutes of every session on a relaxation programme. Examination syllabuses, backed by parents and teachers who demand speed, conspire to frustrate proper learning. Some of my own most rewarding teaching was with two very slow junior children in a special unit. For two separate hours every week for a term I had a completely free hand. The results were remarkable, but increasing cuts and lack of resources in schools mean that such ideal conditions are becoming ever more rare.

Not all mathephobia is avoidable. Laurie Buxton believes that recall of a piece of learning which took place at the time of a traumatic unconnected event in a person's life can bring to the surface emotional responses to the event. These can be so distressing as to convert subconsciously unwillingness to recall the mathematics into inability to do so. A person whose parent died while they were learning division might be for ever incapable of dividing.

Avoidance is the most common way for children to cope with mathephobia. They prefer to sit in silence rather than ask questions and risk ridicule. Others play the clown or affect a lack of concern and pride at their inability to do mathematics. Does 'I can't do mathematics' really mean 'I think I might not be able to do it so I shan't risk trying'?

There is another mathematical fear which is utterly different from mathephobia. This is fear of success, occurring mostly in high ability girls (Leder, 1982). They fear social rejection and loss of self-esteem from doing well in a traditionally male-dominated subject.

Struggle is not always to be avoided, and confidence is increased by overcoming obstacles, not by removing them. However, it should be 'disconcerting for mathematicians to realize the extent of the antipathy to the subject among the population' (Sewell, 1981). The irregularities of our spelling and pronunciation, incomparably more illogical than mathematics, make learning to read in English an abominable task. Yet most people manage it. I find it very sad that something which gives me so much pleasure should be so distressing to others, and know I have been lucky to teach in conditions incomparably easier than a classroom, where it is easier to avoid and overcome mathephobia. The present political climate favours a return to Victorian educational values, the beliefs of Isaac Todhunter and an atmosphere which nourishes mathephobia. I am grateful to Myrtle who gave me the opportunity to 'teach some mathematics to an ardent hater of the subject' (Buxton, 1981). I hope I am wrong, but I fear there may be many more such opportunities in the future.

References

AIKEN, R. (1970) 'Attitudes towards Mathematics', *Review of Educational Research*, 40.

AUDEN, W.H. (1973) *The Observer*, 20 May.

BUXTON, L. (1981) *Do You Panic about Maths?*, London, Heinemann Educational Books.

GRIFFITHS, H.B. and HOWSON, A.G. (1974) *Mathematics: Society and Curriculum*, Cambridge, Cambridge University Press.

HERITAGE, R. (1977) *Mathematical Education for Teaching*, 3, 1. pp. 9–12.

HOLT, J. (1969) *How Children Fail*, Harmondsworth, Penguin.

JOFFE, L.S. (1981) *Dyslexia and Attainment in School Mathematics*, Unpublished PhD thesis, University of Aston.

LEDER, G. (1982) 'Mathematics Achievement and Fear of Success', *Journal for Research in Mathematics Education*, 13, 2, pp. 110–23.

MAXWELL, J. (1984) *Is Mathematics Education Politically Neutral*?, Unpublished MEd dissertation, University of Birmingham.

MORRIS, J. (1981) 'Math Anxiety: Teaching to Avoid It', *Mathematics Teacher*, 74, 6, pp. 413–17.

RESEK, D. and RUPLEY, W. (1980) 'Combatting Mathephobia', *Educational Studies in Mathematics*, 11, 4, pp. 423–41.

SEWELL, B. (1981) *The Mathematical Needs of Adults in Daily Life*, London, ACACE.

MULTICULTURAL AND ANTI-RACIST MATHEMATICS

The previous section looked at a controversial area in the teaching of mathematics — the issue of social and political values. Another contentious area is the movement for anti-racist mathematics, founded by a number of mathematics teachers and educators concerned with issues of race and multiculturalism. For example, the Campaign for Anti-Racist Mathematics was active in Lambeth Teachers Centre, London, in 1986. This movement was attacked by the Prime Minister, Margaret Thatcher, who told her party conference in October 1987 that: 'children who need to be able to count and multiply are learning anti-racist mathematics — whatever that means — and political slogans' (*The Guardian*, 3 November 1987).

Anti-racist mathematics, in fact, merely consists of a multicultural approach to mathematics which is particularly vigilant on matters of racism. Although it has been portrayed as extreme in the media, many of those who are aware of the extent of racism in Britain understand and sympathize with the concern to ensure that the teaching of mathematics is not the bearer of unconscious prejudice. However, there are further issues involved, which are worth making explicit.

First of all, there is the myth of monocultural Britain. Britain is rich in the cultures of immigrant communities. The best known are black cultures, both Afro-Caribbean and Asian. But there are also Chinese, Arab, Greek, Turkish, Polish, Hungarian, Italian, Irish and all sorts of other immigrant communities (increasingly British born) adding to the cultural wealth of these isles. Ignoring all of these groups, Britain is still rich in its cultural diversity. As well as being made up of four countries with three or four different languages, there are regional cultural variations, dialects, even languages (Cornish, for example). In Northumberland shepherds still sometimes count sheep Yan, Tan, Tethera, Methera, Pimp, . . . so there are even cultural variations in mathematics — in number names — in Britain. There are variations in religion, traditions and many other aspects of life. Historically Britain has been populated by

Ancient Britons, Phoenicians, Romans, Vikings, Angles, Saxons, Normans, not to mention having mixed blood with the French, Dutch, Germans, Spanish, and so on. Throughout its history Britain has been one of the most mixed of all countries — in terms of races, countries of origin, culture, and so on. So much for the myth of monocultural Britain. It always has been a multicultural society in reality, and it is about time we recognized this in school in mathematics lessons in particular. There is no unique, traditional British culture which might be compromised by a multicultural approach to education, especially not in mathematics, since it owes so much to other cultures.

Secondly, there is the cultural basis and foundation of mathematics itself. Since mathematics is a dynamic, living, cultural product (as argued in the introduction to this part of the book), the school curriculum must reflect this. Mathematics needs to be studied in living contexts which are meaningful and relevant to the learners. Such contexts include the languages and cultures of the learners, their everyday lives, as well as their school-based experiences. If mathematics is to empower learners to become active and confident problem-solvers, they need to experience a human mathematics which they can make their own.

The three papers in this section explore aspects of these issues: the nature of multicultural and anti-racist mathematics, the issues raised, and finally one avenue to humanizing mathematics, the use of history.

20 *Multicultural and Anti-Racist Mathematics Teaching*

Derek Woodrow

It is a commonly held belief that mathematics is an essentially acultural subject. It is arguable whether this is a valid statement — the nature of argument and the language of implication are both culturally determined — but it is certainly not true that the teaching of mathematics can be acultural. The attempt to convey the ideas and concepts to the learner must take place using the metaphors and imagery available to the learner, and these are clearly the consequence of the society and culture within which the learner lives. Bishop (1985, 1986) has raised the issue of the social construction of meaning in mathematics, and reports that even simple actions such as counting on the fingers appear to have regional variations. He points to many further examples of cultural influences in significant mathematical conceptual structures (see also Lea. 1987). D'Ambrosio (1985) argues that the very learning of mathematics creates a conflict with the 'natural' or 'folk' processes which the learner might utilize; he calls these ethnomathematics. The learning of a formalized response to a situation prevents the application of the more free-wheeling methods. Thus a partially learned (or inadequately understood) procedure invoked by learning mathematics may well be less effective in practice than an informal, intuitive (even if less powerful) method. It is not uncommon for pupils to produce intuitive, accurate responses to calculations but be quite unable to apply the usual algorithms.

Within a mass education system it is perhaps arguable that it is not even appropriate for any subject to attempt to avoid cultural implications. In such a system any element within the curriculum must be principally directed towards the aims and objectives of education, which are concerned with the development of effective and participatory citizens committed to the support and development of their society (see Ernest, 1986). Mathematics exists within the compulsory curriculum because it is effective in educating pupils; the creation of mathematicians, whilst important, is very much a secondary consideration. In this context the comment reported in the Cockcroft Report (*Mathematics Counts*, 1982) that 'many lessons in secondary schools are very often not about anything' becomes a significant and important concern. To quote from Cuff and Payne (1984), 'Mathematics Counts — but not for all' is only acceptable in a selective and option situation, and not in a mathematics-for-all curriculum. In a compulsory curriculum it is not acceptable to dismiss pupils as 'not able to do mathematics' without thereby identifying a special educational need for which particular provision must be made. This need for mathematics education to relate to people and society was anticipated some years ago by David Wheeler in a lecture to the Association of Teachers of Mathematics. He challenged the annual conference that for ten years the final lecture should be concerned with 'humanizing' mathematics education. Society has raised the issues of racism and the development of a genuinely pluralist society as major and significant concerns. Education has consequently been charged with working towards this as a priority concern, and mathematics teachers have begun to ask how they can respond.

In responding to these changing priorities within society and education it is important to notice that the issues which arise are very often already significant concerns of mathematics education. The response to the multicultural society in effect resonates with those developments and concerns already identified: pupil-centred learning styles, the need to develop ideas from those which pupils already hold, active rather than passive learning, problem-centred objectives, group work and methods of using the power of discussion. In looking at the issues which consideration of multicultural/anti-racist concerns have raised we are in effect reviewing the significant current concerns of mathematics teaching in general.

Reflecting Other Cultures

It is symptomatic of the philosophical position held by mathematics teachers that when the issue of multiculturalism first arose the initial search was for mathematical content items. Hemmings (1980a, 1980b, 1984) offered an excellent collection of such items, including the very rich field of Islamic art patterns (see also Tahta, 1987). The impact of this work in creating effective internal imagery can also be seen influencing the work of the Leapfrogs Group (published by Tarquin Products) and in the very evocative collages frequently found in recent issues of *Mathematics Teaching*, edited by Hemmings and Tahta. Zaslavski (1975, 1979) introduces similar activities from an African context, and it is interesting to see almost the same activity appearing as a Rangoli pattern in Hemmings, a unicursal curve in Zaslavski and in a microcomputer activity and investigation called 'snook' (all three involve an $(n \times m)$ lattice). Joseph (1984) also suggests some fascinating methods of calculating using a system called Vedic arithmetic which offers a quite different approach to algorithms. Another useful activity is the investigation of different calendar patterns, our own sun-dominated Roman system compared with the Islamic moon-focused year (when will Ramadan fall near the shortest day in our year — why should that be important?) and the contrasting methodology of our leap year with the Jewish extra month.

All these topics provide excellent opportunities for mathematical activity and belong in the mathematics curriculum without requiring the justification of multicultural concerns. Which multicultural education objectives, however, are attained by such insertions? It is important not to import content merely in order to satisfy external requests and pressures. Tokenistic responses never meet such needs and usually result in increasingly heightened concern. Care must also be taken not to introduce such topics as marginal and trivial activities since this can imply a dismissive view of other societies and values. This same problem relates to the inclusion of historical information in an attempt to counter the Eurocentric view commonly held of the development of mathematics. Most English people tend to think that mathematics really started with the Greeks and gradually worked its way north. Whilst they might acknowledge the very early contribution of the Chinese and Indian mathematicians, the abiding thought is that that was a very long time ago, with perhaps an implicit unspoken question as to what they have done since then. Such chauvinistic attitudes are at the very centre of the issue of multicultural and anti-racist mathematics, and concern the views and attitudes of teachers rather than pupils. The transmission and continuity of such intuitive responses through the classroom make the issue so vital.

The significance of including such elements in the curriculum is also affected by the structure of the community within which the education is taking place. In a culturally mixed classroom it is important that pupils feel able to bring their own cultures into the classroom and that those cultures are acknowledged and respected. This echoes the growing commitment of mathematics teaching to a more pupil-centred action curriculum and towards a constructivist view of learning. This is particularly important with younger children who are less likely to impose their own images on those presented. In presenting pictures, images and illustrative references it is important that teachers should not present only those which come from their own experience. Where a classroom is monocultural, the significance of presenting alternative images takes on a quite different, though no less important, aspect. It is particularly in such classrooms that open and accepting attitudes to alternative ideas and customs need to be fostered and encouraged. It may happen that such situations give rise to racist and discriminatory comments and attitudes. It is important that schools have policies for meeting such eventualities, since without such professional support teachers may seek to suppress rather than to expose views which unless made public are difficult to change.

Educating for a Multicultural Society

It is an implicit assumption of education that one of the most significant ways of affecting attitudes and beliefs is through the acquisition of knowledge. It is for this reason that the often quoted comment that 'mathematics lessons are not about anything' is so important. Gerdes (1981), in describing mathematics education in Mozambique, indicates the ways in which less politically inhibited societies attempt to influence their future citizens, and in the analysis implies that all mathematics curricula carry such elements whether explicit and deliberate or implicit and involuntary. The distinction between 'folk' mathematics [sic] and colonial mathematics as compared to 'democratic' mathematics bears resemblance to changes in English school mathematics from what we might term the 'clerical' mathematics in Victorian schooling to 'industrial' mathematics in the first part of this century to the mathematics of understanding and processes which dominates current discussions. Remnants of previous ages always remain and mathematics (rules without reason) have not disappeared from most syllabuses. Less acceptable in the English context, however, is the intrusion of explicitly value-laden extra-mathematical concepts, such as Gerdes' 'guerrilla fighters', which will be seen as bordering upon 'indoctrination', although the promotion of literacy and health care by such means is probably less objectionable. There is growing evidence of the ways in which similar but far less obvious indoctrination takes place within English texts (see, for example, ILEA, 1986; Northam, 1982). Other examples of implicit value judgments are given by Maxwell (1985), who also intimates that since such implications are unavoidable they are best planned and intended. This is particularly significant for mathematics, which is assumed to be free of value judgments and social commentary, and the indoctrination is therefore in some senses subliminal and less susceptible to rejection. The uncommented incorporation of discussion of stocks and shares in mathematics texts assumes that such activities are legitimate, which they are, of course, in a free Western economy. The implied assumption that women should earn less than men, common in many questions about wages, is not supportable in the same way, and some examination boards now take care to 'balance' such questions so as to avoid reinforcing any latent assumptions. The humourist Stephen Leacock long ago pointed up the assumptions about the virtues of hard work contained in the stories of A, B, and C digging ditches which often provided the meaning of 'problems' in the texts of the 1940s and 1950s. Values and judgments with which a society is in total agreement are often invisible to readers from that society who assume that such inclusions merely represent the natural order of things. Where groups within the society do not concur, however, then the inclusion of debatable issues often causes controversy, and ultimately censorship. The inclusion in a GCE mathematics examination of a question relating the cost of feeding the world to the cost of nuclear weaponry resulted in the imposition of a vetting committee for future papers, with an assumption that either such content had arisen by accident or that views of one group of examiners should outweigh the views of another group.

Less controversially the mathematics curriculum could make a positive contribution to the provision of information about our society in all its facets including those relevant to its multicultural nature. Grugnetti (1979) in a fascinatingly titled article, 'School Mathematics Makes Sardinians Healthier', provides an example of the deliberate insertion of information valuable to pupils, in this case a statistical investigation of hereditary diseases likely to afflict the pupils. The increasing use of problem-solving within the mathematics curriculum will support such a developing programme of conveying useful general information. It should also lead to a clearer sense of relevance for pupils of the mathematics they are studying. But it may well depend upon the ability (willingness?) of teachers to abandon prescribed plans on occasion in order to follow up topical events (the question of what constitutes an uneconomic pit is a very mathematical concern but would obviously have needed a very sensitive, and probably whole school, approach). Health issues are of interest for insertion into the curriculum for social rather than mathematical reasons, but it is also possible to provide situations in which different eating patterns are made evident and in which comparative costs in different contexts can be investigated; an interesting analysis has been made of comparative costs of items in 1935 and 1985, using the cost of a Mars bar as the unit of value. The reason why some societies do not drink coffee or eat cornflakes can soon be made obvious by reference to their comparative costs. For many pupils the enormous variety of vegetables and foods provides a much wider experience than was available in past times, and enables many issues to be raised such as their costs, including the transport and handling charges needed to provide the rapid delivery some need. For some pupils there may be a need deliberately to widen their awareness of the way in which the international market operates. There are a number of sources of useful information, in particular the Development Education Centres, and some examples of material in use will be found in an article by Hudson (1987), who has also produced a database of information. As indicated above with the question about uneconomic pits there are many pitfalls for the unpractised mathematics teacher who may be unwittingly drawn into issues of opinion and belief, and the experience of colleagues in other disciplines more used to facing this dilemma may need to be utilized. On many social issues the teacher's own beliefs may need to be subsidiary to those of the society he/she is called upon to represent.

Learning by Individuals

At all levels of education there exists a conflict between learning as a social activity and the individual level at which it occurs. Whilst the problem may be the same, however, the symptoms and solutions may be very different. In the early years the issue of how and when to use the mother tongue is a concern, and appropriate materials may be difficult to come by unless the support of the home and the involvement of parents are sought. There is then the problem of the second-hand transmission of what may be subtle ideas (see Dawe, 1983; Jones, 1982), nevertheless the creation of strong home/school identity is likely to be indispensable for maintaining a harmonious society. Emblem (1986) presents a valuable picture of issues in teaching young children, unusually well informed by a visit to the Indian subcontinent. Some learned skills such as visual discrimination seem to be very dependent on the ambient culture (see Mitchelmore, 1980), and although these are likely to be less noticeable in a single dominant environment, it is an area in which pupils might be expected to show wide variations as a result of varying experiences. There is also the issue of styles of learning, which whilst affected by many ingredients, is also related to home culture and such issues as the nature of authority and adult-child relationships (see Head, 1981). Whilst these issues do not directly affect the mathematics which is taught, they are major elements in the decisions about how it is taught, and the content and the teaching approach have a symbiotic relationship. Thus rule-bound algorithmic skills require an essentially authoritarian environment, whilst real problem-solving may well require an act of deliberate withdrawal of authority by the teacher. Pupils inevitably respond differently to these situations, just as some are reticent and others forthcoming, some are born to be leaders whilst others have leadership thrust upon them. These differences exist across and within all cultures and all races but the ways of responding and dealing with these differences are often cultural traits.

It was stated earlier that most of the issues raised by the multicultural debate in mathematics teaching were already concerns of such professionals. The provision of a more accepting, acceptable and enjoyable mathematical environment for all pupils would help towards the achievement of a more accepting, acceptable and enjoyable society.

References

Bishop, A.J. (1985) 'The Social Construction of Meaning — a Significant Development for Mathematics Education?' *For the Learning of Mathematics*, 5, 1, pp. 24–8.

Bishop, A.J. (1986) 'Mathematics Education as Cultural Induction', *Nieuwe Wiskrant*, October, pp. 27–32.

Cockcroft, W.H. (1982) *Mathematics Counts*, London, HMSO.

Cuff, E.C. and Payne, G. (1984) *Crisis in the Curriculum*, London, Croom Helm.

D'Ambrosio, U. (1985a) 'Ethnomathematics and Its Place in the History and Pedagogy of Mathematics', *For the Learning of Mathematics*, 5, 1, pp. 44–8.

D'Ambrosio, U. (1985b) 'Mathematical Education in a Cultural Setting', *International Journal of Mathematics Education in Science and Technology*, 16, 4, pp. 469–77.

Dawe, L.C.S. (1983) 'Bilingualism and Mathematical Reasoning in English as a Second Language', *Educational Studies in Mathematics*, 14, pp. 325–52.

Emblem, V. (1986) 'Asian Children in Schools', *Mathematics in School*, 15, 5, pp. 26–9; 16, 1, pp. 7–9.

Ernest, P. (1986) 'Social and Political Values in Mathematics', *Mathematics Teaching*, 116, pp. 16–18 (reprinted as Chapter 17 in this book).

Gerdes, P. (1981) 'Changing Mathematics Education in Mozambique', *Educational Studies in Mathematics*, 12, pp. 455–77.

Grugnetti, L. (1979) 'School Mathematics Makes Sardinians Healthier', *Mathematics in School*, 8, 5, pp. 22–3.

Head, J. (1981) 'Personality and Learning of Mathematics', *Educational Studies in Mathematics*, 12, pp. 339–50.

Hemmings, R. (1980a) 'Multi-Ethnic Mathematics: Part I: Primary', *New Approaches to Multicultural Education*, 8, 3, pp. 29–38.

Hemmings, R. (1980b), 'Multi-Ethnic Mathematics: Part II: Secondary', *New Approaches to Multicultural Education*, 8, 4, pp. 1–4.

Hemmings, R. (1984) 'Mathematics', in A. Craft and G. Bardell (Eds), *Curriculum Opportunities in a Multicultural Society*, London, Harper-Row.

Hudson, B. (1987) 'Global and Multicultural Issues', *Mathematics Teaching*, 119, pp. 52–5.

ILEA (1986) *Everybody Counts*, London, ILEA.

Jones, P.L. (1982), 'Learning Mathematics in a Second Language: A Problem with More and Less', *Educational Studies in Mathematics*, 13, pp. 269–87.

Joseph, G. (1984) 'The Multicultural Dimension', *The Times Educational Supplement*, 5 October.

Lea, H. (1987) 'Traditional Mathematics in Botswana', *Mathematics Teaching*, 119, pp. 38–41.

Maxwell, J. (1985) 'Hidden Messages', *Mathematics Teaching*, 111, pp. 18–19.

Mitchelmore, M.C. (1980) 'Three-dimensional Geometric Drawing in Three Cultures', *Educational Studies in Mathematics*, 11, pp. 205–16.

Northem, J. (1982) 'Girls and Boys in Primary Mathematics Books', *Education*, 10, 1, pp. 3–13.

Tahta, D. (1987) 'Islamic Patterns', *Mathematics Teaching*, 119, pp. 48–51.

Zaslavski, C. (1975) 'African Network Patterns', *Mathematics Teaching*, 73, pp. 12–13.

Zaslavski, C. (1979) *Africa Counts*, London, Lawrence Hill.

21 *What Is Multicultural Mathematics?*

Marilyn Nickson

Classroom mathematics has undergone many changes throughout recent decades, each of which has been brought about by a variety of agencies and has exerted a new kind of demand on the curriculum. In recent years the Cockcroft Report (1982) has called into question what was seen as a back-to-basics movement at primary level, and at secondary level has made recommendations which have influenced the formulation of a mathematics curriculum leading towards the new GCSE examination. These changes have been brought about to a great extent in response to society's perceptions of what is seen as failure on the part of schools to teach mathematics successfully enough to meet the demands of a highly technological society. In recent years a new concern has surfaced which adds a further perspective to the problems which influence the effective teaching and learning of the subject. This concern arises from the extent to which society has increased in cultural diversity and the way in which resultant cultural differences have come to manifest themselves more strongly in schools. A key problem now being faced is how accommodate these differences within the curriculum so that the learning of mathematics can be enhanced for all.

There are essentially two points of view which provide the ends of a spectrum of approaches along which such accommodation might be achieved. A brief exploration of each of these helps to clarify the different interpretations of multicultural mathematics and to shed some light upon how they can affect classroom practice.

Mathematics as 'Culture-Free' Content

Any consideration of how to accommodate cultural diversity within the mathematics curriculum presupposes that this is a feasible undertaking in the first place. However, the nature of mathematics as it is perceived by many suggests that it is 'culture-free' and need not (indeed, cannot in the view of some) be 'adapted' to the idiosyncrasies which may be determined by the culture of an individual or individuals. This view of the nature of mathematics is one which has long been held and probably still is dominant in the minds of many teachers and teacher educators, and tends to be reflected in the pedagogies that they adopt. Studies and surveys of mathematics education in recent decades at primary and secondary level have tended to reflect a curriculum in both content and style of teaching that is dominated by this kind of perspective (e.g., Cockcroft, 1982; DES, 1978, 1979; Hilsum and Strong, 1978; Ward, 1979). In spite of the efforts of some, (including professional bodies such as the Association of Teachers of Mathematics and the Mathematical Association) to bring about change through the adoption of different methodologies and approaches to content, the influence of this kind of belief about the nature of mathematics is still strong. While it is clear that new demands such as those being imposed by the GCSE involving a more open-ended approach with investigational, practical and oral work could, with time, gradually lead to the questioning of long-held beliefs about mathematics, such moves are already being seen to have a somewhat superficial influence. For example, there are signs that investigational work is becoming rigidified rather than providing an opportunity for a more exploratory approach to mathematical ideas as was intended (see Chapter 6). Taken together, attempts to free teachers from this rigid perspective have not met with a great deal of success so far.

The epistemological foundations that have led to mathematics being accepted as unchallengeable, received knowledge lie in what Lakatos (1976) calls a 'formalist' view of the subject firmly grounded in logical positivism. Many in mathematics education have been transfixed by this formalist perspective of the subject which, as Lakatos (1976) suggests, allows a view of mathematics in terms of formal systems to dominate at the expense of a more open and exploratory interpretation. He describes those who hold such a view as 'dogmatists' who see mathematical ideas as having been 'purged of all the impurities of earthly uncertainty' (Lakatos, 1976, p. 2).

It could be particularly difficult for anyone under the influence of such a view of the subject to allow that uncertainties may arise as a result of the differing cultural backgrounds of individuals engaged in mathematical activity. If mathematical knowledge is by definition pure and unadulterated, then the task is to see that it is received as such and that it should not be tampered with. This in turn suggests that any intervention in the mathematics curriculum to accommodate cultural differences of learners (if it takes place) is unlikely to be of a very fundamental nature; on the contrary, it will necessarily be of a superficial kind.

This approach can be seen in attempts to adopt a multicultural approach to the curriculum which in essence set out to provide different cultural contexts for traditional types of mathematical problem. The words change but methodology and content remain the same. Another approach is to introduce more of the history of mathematics, and thus to show its multicultural nature by identifying the important roles played by the Chinese, Indians and Mayans, for example, and not to leap in time from the Greeks to Western culture in the development of mathematics as we know it now (see, for example, Hudson, 1987). In a sense such approaches are accommodating the existence of other cultures by referring to different cultural artefacts but not by considering the possible cultural differences that exist in ways of thinking or construing that have brought such artefacts about. This leads us to consider the other end of the spectrum of approaches to multicultural mathematics referred to earlier.

Mathematics as a 'Culture-Bound' Way of Thinking

The extreme opposite to the notion of culture-free mathematics would be that of mathematics as culture-bound. This immediately suggests a number of 'different' bodies of mathematical knowledge that could be mutually exclusive and contradictory. The idea of the total objectivity of mathematical thought would seem to disappear together with the notion of absolute truth linked with it. How can such a viewpoint be accommodated and, in particular, how can contradictory ideas exist side by side?

The answers to such questions lie, in the first place, in adopting a different view of the nature of mathematical knowledge and how it comes into being. The essential difference from that of the formalist perspective lies in accepting the fact that mathematics is founded in social activity and human intercourse as is any other kind of knowledge.

It is no longer accepted as infallible knowledge generated in some erudite vacuum which, as Lakatos (1976) suggests, 'denies the status of mathematics to most of what has been commonly understood as mathematics' (p. 2). Contradictory ideas become competing theories as mathematical knowledge develops. Objectivity is arrived at, as Toulmin (1972) says, by taking into account the accumulation of experiences 'in all cultures and historical periods', thereby reaching an objective point of view 'in the sense of being neutral' (Toulmin, 1972, p. 50). The adoption of such a fallibilist approach to the nature of mathematics permits us to approach mathematical knowledge and to interpret it in terms of its growth and context in which it grows and has grown.

The adoption of a perspective of mathematics seen in these terms can have a profound effect on how the subject is approached in the classroom, and several mathematics educators have explored the potential outcomes and benefits of such an approach (e.g., Confrey, 1980; Nickson, 1981; Wolfson, 1981; Pimm, 1982; Lerman, 1986). More particularly, however, this theoretical perspective provides a fresh viewpoint from which to approach the whole question of a multicultural mathematics curriculum.

Inherent in the acceptance of the notion of socially constructed mathematical knowledge there is the freedom to recognize cultural influences both at the broad societal level and at the level of the individual. Rather than simply looking for aspects of different cultures to exemplify mathematical ideas (thus supposedly providing familiar social contexts for learners), what we should be looking for is how their cultural perspective may affect their mathematical mode of thought. This is being done at a national level, for example, in Mozambique where the mathematics embedded in local cultural activities (such as basket-weaving) is used as the foundation for curricular mathematics (Gerdes, 1986). However, catering for cultural perspectives in a situation where there may be an assumed cultural and ethnic homogeneity is a different matter from catering for it in a society where this does not exist. It is a tall enough order in the first situation, but it becomes an even taller order if we consider culture as, for example, D'Ambrosio (1986) does when he writes: 'We have built a concept of society out of cultural attitudes and cultural diversity, that is, different groups of individuals behave in a similar way, because of their modes of thought, jargon, codes, interests, motivation, myths' (p. 5).

To view culture defined according to such a variety of criteria appears to make a complex problem seem even more complex, but is this, on closer examination, really the case? Defining culture in terms of 'groups of individuals' bound together by factors other than ethnic origin redirects our attention to the individuals *in their social context* and emphasizes the problem of catering for cultural individuality in a shared setting. Where mathematics in particular is concerned, Bloor's (1976) identification of what he calls 'social causes of mathematical thought' provides some indication of the kinds of factors that could be taken into account when trying to accommodate such cultural individuality within the construction of a mathematics curriculum.

Conclusion

It is clear from this discussion that the notion of multicultural mathematics is a very demanding one when it comes to catering for it within school curricula. On the one hand, there is the temptation to opt for a somewhat simplistic solution to the problem by including references to cultural phenomena related to different ethnic groups in teaching and learning materials. This is, as we have seen, often the outcome of a particular set of beliefs about the nature of mathematics which exerts powerful constraints upon how mathematics is done in the classroom. On the other hand, a more complex but more meaningful solution could lie in re-examining those beliefs and in considering mathematical knowledge more in terms of its social nature and foundations. This would help to open up the subject so that the cultural identity of the learner could potentially be more readily accommodated. The freedom to see mathematical knowledge in more problematic terms brings with it a flexibility that can have profound effects upon the way it is approached in the classroom.

It was stated at the outset that these points of view present the extreme ends of a spectrum. Clearly a solution to the problems posed in devising a multicultural mathematics curriculum must lie somewhere between the two and involve a combination of elements of both. In the past there has been an imbalance in favour of the first of these extremes. Until more thought is given to the social nature of mathematical knowledge and its implications for the curriculum we are unlikely to succeed in devising a mathematics curriculum that meets the demands of the multicultural society in which we live.

References

Bloor, D. (1976) *Knowledge and Social Imagery*, London, Routledge and Kegan Paul.

Cockcroft, W.H. (1982) *Mathematics Counts*, London, HMSO.

Confrey, J. (1980) 'Conceptual Change Analysis: Implications for Mathematics and Curriculum Inquiry', Paper presented at the AERA National Meeting, Boston, February.

D'Ambrosio, U. (1986) 'Socio-Cultural Bases for Mathematical Education', in M. Carss (Ed.), *Proceedings of the Fifth International Congress on Mathematical Education*, Boston, Mass., Birkhauser.

Department of Education and Science (1978) *Primary Education in England: A Survey by HMI Inspectorate*, London, HMSO.

Department of Education and Science (1979) *Aspects of Secondary Education*, London, HMSO.

Gerdes, P. (1986) 'On Culture: Mathematics and Curriculum Development in Mozambique', in H.M. Johnsen and S. Mellin-Olsen (Eds), *Mathematics and Culture*, Bergen, Caspar Forlag, Bergen Laererhogskole.

Hilsum, S. and Strong, C. (1978) *The Secondary Teacher's Day*, Slough, NFER.

Hudson, B. (1987) 'Multicultural Mathematics', *Mathematics in School*, 16, 4, pp. 34–8.

Lakatos, I. (1976) *Proofs and Refutations*, Cambridge, Cambridge University Press.

Lerman, S. (1986) *Alternative Views of the Nature of Mathematics and Their Possible Influence on the Teaching of Mathematics*, Unpublished doctoral dissertation, King's College, University of London.

Nickson, M. (1981) *Social Foundations of the Mathematics Curriculum: A Rationale for Change*, Unpublished doctoral dissertation, University of London Institute of Education.

Pimm, D. (1982) 'Why the History and Philosophy of Mathematics Should Not Be Rated 'X'', *For the Learning of Mathematics*, 3, 1, pp. 12–15.

Toulmin, S. (1972) *Human Understanding*, Princeton, N.J., Princeton University Press.

Ward, M. (1979) *Mathematics and the Ten-Year Old* (Schools Council Working Paper 61), London, Evans/Methuen Educational.

Wolfson, P. (1981) 'Philosophy Enters the Mathematics Classroom', *For the Learning of Mathematics*, 2, pp. 22–6.

22 *The Use of the History of Mathematics in Teaching*

Derek Stander

Life without some knowledge of mathematics would be very difficult in the latter part of the twentieth century. In some cases mathematics is essential for continuing one's profession or to understand one's personal affairs. Many young people are put off mathematics either by bad teaching or by being exposed to too much mathematics at an early stage in their education. The opinion has been expressed by some college lecturers and teachers that the content of the new GCSE mathematics syllabus is even now too large.

Motivation and interest may be promoted in mathematics by showing the practical value of learning the subject. Experimental and discovery work can also be employed to raise the pupil's interest. The history of mathematics can be used to provide some enrichment of the subject and present a view that mathematics is connected with the development of our culture. Simons (1923), writing in the *Mathematics Teacher*, suggested that using history and recreations was a way of vitalizing the teaching of mathematics. This opinion was supported by Hassler (1929), writing in the same journal. Hadamard (1954) says that Henri Poincaré advocated an historical approach to mathematics teaching. Barzum (1947) and Wiltshire (1930) express similar opinions to those of Simons and Hassler. The American mathematical historian Jones (1957) strongly encourages the use of the history of mathematics as a teaching tool. From a pupil's point of view it is to be hoped that an historical approach would increase motivation by making the subject more human and by showing that the greatest mathematicians were human beings with human frailties who worked very hard to achieve their results. The pupil's confidence increases with the realization that the great did not write down their answers immediately. An historical approach to mathematics teaching can also help the pupil realize that mathematics is not a once and for all discovery, but is constantly changing. There are links between mathematics and other subjects which can easily be shown to exist by studying the historical development of the subject. It is possible that an historical approach encourages competence as there is greater exposure to mathematics and mathematical ideas. For some pupils a study of the history of mathematics will show that good notation and setting out will help in producing not only the right answer but also many discoveries.

There appear to have been very few previous investigations of any depth into the use of the history of mathematics in the classroom. The only two found by the author were made by D.R. Green (1974) and L.F. Rogers (1976). In a survey conducted by the author no adverse opinion was found for the use of the historical development of topics as a tool in mathematics teaching.

The Use Being Made of the History of Mathematics in Education

In order to find out the use being made of the history of mathematics in education a number of primary schools were visited, a questionnaire was sent to the mathematics advisers of all LEAs in England and a second questionnaire was sent to the mathematics departments of all the universities on the British mainland. From the information gathered about the teaching of mathematics in these establishments is was found that little use was being made of the history of mathematics.

Before this investigation was started it was assumed that in the primary school sector of state education great use would be made of the history of mathematics to enrich mathematical studies as a means of providing such project materials for pupils. The reality is, however, quite different.

In the secondary sector it is important to differentiate between history of mathematics courses and the history of mathematics being used as enrichment or a teaching tool. There are very few history of mathematics courses being taught. Green (1974) reported less than one in twenty of 1 per cent of the total examination entry in mathematics being made up of history of mathematics course candidates. The LEA mathematics advisers in their replies to the questionnaire indicated that there was little use being made of the history of mathematics in secondary schools. They suggested that they knew of a few teachers who told their pupils of the achievements of some named mathematicians, but in general there was no indication of departmental policy to use the historical development of mathematics as a teaching tool.

In the universities there appears to be an underlying desire to make greater use of the history of mathematics in the teaching of mathematics. A few universities offer history of mathematics courses. A lack of courses means a lack of potential lecturers. This in turn leads to a lack of courses — a cycle! It was obvious considering the evidence from lecturers who replied to questionnaires or who were interviewed during the investigation that there is in universities considerable potential for making use of the history of mathematics. One problem, however, appears to be that university lecturers have to include a high mathematical content in their courses, and they have little opportunity either to prepare material which they consider suitable or to present this material when they have prepared it.

It was thought reasonable as part of this investigation to compare the use of the history of mathematics in Britain with its use in other countries. Thirty-one other countries were selected according to their secondary school enrolment ratio. The London embassies of the thirty-one countries were contacted with requests for information about the teaching of mathematics in their countries with particular reference to the history of mathematics. From the replies received it was evident that in the Scandinavian countries, Denmark and Norway, considerable interest is taken in the history of mathematics. The history of mathematics is much used, investigated and taught in the United States of America, which is leading the field in this aspect of mathematics. Little evidence was found that any other country used the history of mathematics.

Are Attitudes towards Mathematics Changed by Introducing the History of Mathematics to the Mathematics Classroom?

Two experiments were carried out to study changes in attitude towards mathematics when history was used in the teaching of mathematics. These experiments involved pupils from secondary schools and student primary school teachers in training. An attitude questionnaire taken from Shaw and Wright (1967) consisting of twenty items using the Likert scaling procedure was used. This attitude questionnaire was designed to measure changes in attitude when the history of mathematics was used in teaching mathematics.

In the first experiment the mathematics and history of the Euler relationship ($V - E + F = 2$) for convex polyhedra were used. The experiment was carried out with twenty-two girls in a private school and forty-one boys and girls in a comprehensive school. In each school the pupils were divided into two groups. One group was presented with a teaching package containing the mathematical material connected with the Euler relation. This group was regarded as a control group. The second group was presented with the same mathematical material but with additional material which was concerned with the discovery and proof of the Euler relationship. Each group took two weeks to study these materials. An analysis of the results obtained from the two schools showed that there was no change in attitude within or between the history group and the control group in the private school. Similarly there was no significant change in attitude in either of the groups in the comprehensive school, and again there was no significant difference between the groups.

This experiment clearly showed that a limited short-term inclusion of some history of mathematics material had no effect on pupil attitudes towards mathematics. Although it is well known that attitudes can be changed, any change might take some time to effect. This is possibly the case when history is introduced into mathematics teaching.

The second experiment involved potential primary school teachers. The entry of students to a university School of Education was divided into two groups. One of the groups was given a weekly historical enrichment sheet about the mathematics they had been studying. This was done for one term. The second group or control group was given no enrichment material. A Likert-type attitude questionnaire was given to all the students before the experiment was begun. The same questionnaire was again given at the end of the experiment. An analysis of the results showed that using this kind of enrichment material made no difference to the attitudes of these future teachers. This might be because of the limited time devoted to this experiment. It is worth noting here that when these future teachers were interviewed, the majority stated that they enjoyed reading about the mathematicians connected with the mathematics they were studying, but that they would be reluctant to spend time finding out about the historical development for themselves.

History of Mathematics Resources

Resource materials for the history of mathematics can be found in books and mathematical journals. Films and videos, sound tape sequences, filmstrips, single slides and wall charts are also available.

The author has identified 493 books about the history of mathematics with 19 per cent of these books being in print in 1986. A book was included if it met two criteria: it had to be written in English, and the major part of its content had to be about the history of mathematics. The four most popular topics written about were numbers, Euclidean geometry, Greek mathematics and astronomy. The five most popular mathematicians were Newton, Galileo, Einstein, Babbage and Kepler. The majority of the books identified are excellent reading for mathematics teachers, but regrettably there are few that would be suitable for pupils up to fifth form level in the secondary school.

Articles about the history of mathematics appear infrequently in mathematical journals, but when they do they are generally of very high quality and are often difficult to read because of the high level of the mathematics being discussed.

Only six films/videos about the history of mathematics have been identified by the author. One of these was a very useful cartoon called

'Donald in Mathmagic Land', which would provide an enjoyable background to geometry and trigonometry teaching. Four of the others were about the history of measurement, and the sixth was a film used to introduce the Open University Mathematics Foundation course.

All the sound tape sequences identified as being currently available are about measurement. There are filmstrips and single slides available particularly from the Bodleian Library, Oxford. (The Bodleian Library will send a detailed catalogue on request.)

IBM produce a free frieze poster which is about twelve feet long and gives a large amount of detail about mathematicians and mathematical events from 1100 AD to the present day. This chart is supplied free to schools and is admirably suitable for classroom display. Posters about the Möbius Band, the Euler Solids and Archimedes are available from the Mathematical Association.

Conclusions

The value of mathematics and the need to learn mathematics are apparent. For learning there generally has to be a teacher. Any teacher has a great opportunity to stimulate interest in his subject. A way that would seem to make mathematics interesting to students and laymen is to approach it in the spirit of enrichment and enjoyment. The application of mathematics to practical problems, however, is a very serious matter indeed. But no student is motivated to learn advanced group theory, for example, if one tells him that he will find it beautiful and stimulating, or even useful, if he becomes a particle physicist. If the teacher uses his time drilling his students in routine operations, he kills their interest, hampers their intellectual development and misuses his opportunity. If he challenges the curiosity of his students by setting them stimulating problems which demonstrate possible applications, and enriching his teaching with the many fascinating aspects of the historical development which shows mathematics as an ever evolving human endeavour, he may induce in them a taste for independent thinking and a liking of mathematics.

The limited experimental evidence available suggests that there is little improvement in attitudes towards mathematics when historical material is included in teaching over a short period of time. To evaluate the effect on attitudes when the historical development of mathematics is employed in teaching would require longer experiments with more pupil involvement. This in turn would necessitate a greater participation on the part of teachers. But teachers are largely untrained in the use or substance of the history of mathematics. Again we are in a cycle. Surely there is a need for one or two schools of education not only to teach history of mathematics courses but also to illustrate to students ways of using history in the mathematics classroom. This would result in a number of mathematics teachers being conversant with the development of their subject. It would then be possible to carry out attitude experiments from a wider base. There might then be a clearer indication of the value of using the historical development of mathematics.

Lack of resources for the pupil certainly inhibits the use of historical material in mathematics classrooms. There is a large volume of historical resource material available for the interested teacher, provided he had available to him a 'friendly' library or Teachers Centre.

An indexed listing of all the books, articles, films/videos, filmstrips/slides, posters and charts and portraits of mathematicians identified by the author about the history of mathematics is available from the Centre for Innovation in Mathematic Teaching, at the School of Education, University of Exeter.

References

Barzum, J. (1947) *Teaching in America*, Boston, Mass., Little, Brown and Company.

Green, D.R. (1974) *The Place of History in the Teaching of Mathematics*, Unpublished MEd thesis, Chelsea College, University of London.

Hadamard, J. (1954) *The Psychology of Invention in the Mathematical Field*, New York, Dover Publications.

Hassler, J. (1929) 'The Use of History of Maths in Teaching', *The Mathematics Teacher*, March.

Jones, P.S. (1957) 'The History of Mathematics as a Teaching Tool', *The Mathematics Teacher*, January.

Rogers, L.F. (1976) *Mathematics, History and Teaching*, Unpublished MEd thesis, University of Leicester.

Shaw, M.E. and Wright, J.M. (1967) *Scales of Measurement of Attitudes*, McGraw Hill.

Simons, L.G. (1923) 'The Place of the History and Recreations of Mathematics in Teaching Algebra and Geometry', *The Mathematics Teacher*, February.

Wiltshire, B. (1930) 'History of Mathematics in the Classroom', *The Mathematics Teacher*, December.

THE MATHEMATICS TEACHER

This volume has explored various themes concerned with the teaching of mathematics. One factor towers above all others in importance: the mathematics teacher, which is the subject of this final section.

In Britain the mathematics teacher has a double importance. Not only is he or she the person who does the teaching, but often he or she also devises the mathematics curriculum. Of course the situation is usually more complex, with the curriculum design initiative resting with a school's head of mathematics, or mathematics coordinator. Within this school management structure and the shared goals and scheme adopted, individual teachers of mathematics have traditionally had a great deal of freedom in planning their teaching of mathematics. Some lack the confidence, knowledge, or whatever it takes to exploit fully this freedom in planning the mathematics teaching for their particular classes, especially in the primary school. Many teachers not only tailor imaginative mathematics curricula to the needs of their classes, but also become involved in designing mathematics schemes for their school, for a regional development, or for a national curriculum project. The key issue raised here is that of teacher autonomy — the extent to which the teacher is self-determining in planning for mathematics teaching. This is treated in the first of the two chapters in this section, which looks to teachers' beliefs and views of the nature of mathematics as a basis for their practices.

The advent of the *National Curriculum* for state schools in Britain in the 1990s may erode some of the autonomy exercised by teachers of mathematics. The imposition of national attainment targets, with standardized testing at the ages of 7, 11, 14 and 16, will clearly restrict the ability of teachers to design and teach mathematics in an imaginative, locally relevant way. Particularly at risk may be the imaginative cross-disciplinary or thematic work in primary schools, in which mathematics may be integrated with science, environmental studies, CDT, art and creative writing, in the exploration of a topic. Such classroom activities are consistent with a creative, problem-solving view of mathematics, and of knowledge-getting overall. However, the way that the *National Curriculum* has reinstated traditional subject boundaries — despite protestations to the contrary — suggests that there will be pressure on teachers to concentrate on more traditional patterns in mathematics teaching.

There is also the possibility that the imposition of the *National Curriculum* may provide some schools with a welcome and useful structure, especially in situations where there is a shortage of adequately prepared teachers of mathematics. This may be just as well. The second chapter in this section treats a number of issues including a concern over the supply of mathematics teachers. As Cockcroft (1982) indicates, there has been concern over the number and quality of mathematics teachers for some time. Unfortunately, for a number of reasons including the demand for mathematics graduates outside teaching and the decline in status and resourcing of education, the shortage of mathematics teachers is growing. Thus although the *National Curriculum* may impinge upon the traditional autonomy of the mathematics teacher, it will do so as the number of teachers capable of functioning autonomously decreases.

There are a great number of issues concerning the teacher of mathematics which inevitably cannot be covered here — even considering issues relating to social context alone. One growth point — a very important one — is research on mathematics teacher education. How mathematics teachers are educated, the roles of reflection, beliefs, knowledge of mathematics, practical knowledge of teaching mathematics and the impact of practical experience and contact with experienced teachers in the development of the mathematics teacher, all of these are beginning to be explored.

Reference

Cockcroft, W.H. (1982) *Mathematics Counts*, London, HMSO.

23 *The Impact of Beliefs on the Teaching of Mathematics*

Paul Ernest

Official reports such as NCTM (1980) *Agenda for Action*, and the Cockcroft Report (1982) recommend the adoption of a problem-solving approach to the teaching of mathematics. Such reforms depend to a large extent on institutional reform: changes in the overall mathematics curriculum. They depend even more essentially on individual teachers changing their approaches to the teaching of mathematics. However, the required changes are unlike those of a skilled machine operative, who can be trained to upgrade to a more advanced lathe, for example. A shift to a problem-solving approach to teaching requires deeper changes. It depends fundamentally on the teacher's system of beliefs, and in particular on the teacher's conception of the nature of mathematics and mental models of teaching and learning mathematics. Teaching reforms cannot take place unless teachers' deeply held beliefs about mathematics and its teaching and learning change. Furthermore, these changes in beliefs are associated with increased reflection and autonomy on the part of the mathematics teacher. Thus the practice of teaching mathematics depends on a number of key elements, most notably:

the teacher's mental contents or schemes, particularly the system of beliefs concerning mathematics and its teaching and learning;
the social context of the teaching situation, particularly the constraints and opportunities it provides; and the teacher's level of thought processes and reflection.

These factors determine the autonomy of the mathematics teacher, and hence also the outcome of teaching innovations — like problem-solving — which depend on teacher autonomy for their successful implementation.

The mathematics teacher's mental contents or schemas include knowledge of mathematics, beliefs concerning mathematics and its teaching and learning, and other factors. Knowledge is important, but it alone is not enough to account for the differences among mathematics teachers. Two teachers can have similar knowledge, but while one teaches mathematics with a problem-solving orientation, the other has a more didactic approach. For this reason the emphasis below is placed on beliefs. The key belief components of the mathematics teacher are the teacher's:

view or conception of the nature of mathematics,
model or view of the nature of mathematics teaching,
model or view of the process of learning mathematics.

The teacher's conception of the nature of mathematics is his or her belief system concerning the nature of mathematics as a whole. Such views form the basis of the philosophy of mathematics, although some teachers' views may not have been elaborated into fully articulated philosophies. Teachers' conceptions of the nature of mathematics by no means have to be consciously held views; rather they may be implicitly held philosophies. The importance for teaching of such views of subject matter has been noted both across a range of subjects and for mathematics in particular (Thom, 1973). Three philosophies are distinguished here because of their observed occurrence in the teaching of mathematics (Thompson, 1984), as well as in the philosophy of mathematics and science.

First of all, there is the instrumentalist view that mathematics is an accumulation of facts, rules and skills to be used in the pursuance of some external end. Thus mathematics is a set of unrelated but utilitarian rules and facts. Secondly, there is the Platonist view of mathematics as a static but unified body of certain knowledge. Mathematics is discovered, not created. Thirdly, there is the problem-solving view of mathematics as a dynamic, continually expanding field of human creation and invention, a cultural product. Mathematics is a process of inquiry and coming to know, not a finished product, for its results remain open to revision.

These three philosophies of mathematics, as psychological systems of belief, can be conjectured to form a hierarchy. Instrumentalism is at the lowest level, involving knowledge of mathematical facts, rules and methods as separate entities. At the next level is the Platonist view of mathematics, involving a global understanding of mathematics as a consistent, connected and objective structure. At the highest level the problem-solving view sees mathematics as a dynamically organized structure located in a social and cultural context.

The model of teaching mathematics is the teacher's conception of the type and range of teaching roles, actions and classroom activities associated with the teaching of mathematics. Many contributing constructs can be specified including unique versus multiple approaches to tasks, and individual versus cooperative teaching approaches. Three different models which can be specified through the teacher's role and intended outcome of instruction are:

Teacher's role	*Intended outcome*
1 Instructor	Skills mastery with correct performance
2 Explainer	Conceptual understanding with unified knowledge
3 Facilitator	Confident problem-posing and problem-solving

The use of curricular materials in mathematics is also of central importance in a model of teaching. Three patterns of use are:

1 the strict following of a text or scheme;
2 modification of the textbook approach, enriched with additional problems and activities;
3 teacher or school construction of the mathematics curriculum.

Closely associated with the above is the teacher's mental model of the learning of mathematics. This consists of the teacher's view of the process of learning mathematics, what behaviours and mental activities are involved on the part of the learner, and what constitute appropriate and prototypical learning activities. Two of the key constructs for these models are: learning as active construction, as opposed to the passive reception of knowledge; the development of autonomy and child interests in mathematics, versus a view of the learner as submissive and compliant. If one uses these key constructs the following simplified models can be sketched, based on the child's:

1 compliant behaviour and mastery of skills model,
2 reception of knowledge model,
3 active construction of understanding model,
4 exploration and autonomous pursuit of own interests model.

Relationships between Beliefs, and Their Impact on Practice

The relationships between teachers' views of the nature of mathematics and their models of its teaching and learning are illustrated in Figure 1. It shows how teachers' views of the nature of mathematics provide a basis for the teachers' mental models of the teaching and learning of mathematics, as indicated by the downward arrows. For example, the instrumental view of mathematics is likely to be associated with the instructor model of teaching, and with the strict following of a text or scheme. It is also likely to be associated with the child's compliant behaviour and mastery of skills model of learning. Similar links can be made between other views and models, for example:

mathematics as a Platonist unified body of knowledge — the teacher as explainer — learning as the reception of knowledge;

Figure 1. Relationships between Beliefs, and Their Impact on Practice

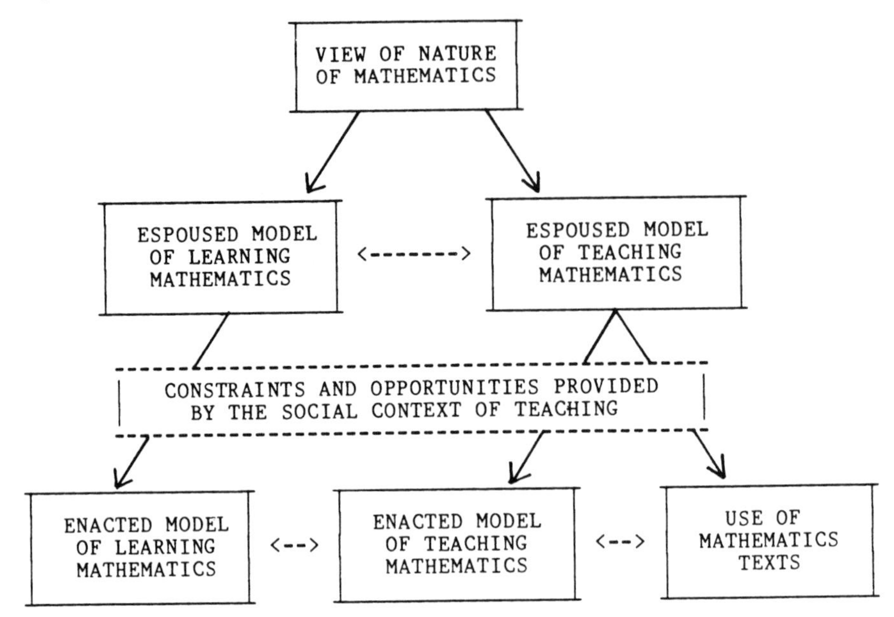

mathematics as problem-solving — the teacher as facilitator — learning as the active construction of understanding, possibly even as autonomous problem-posing and problem-solving.

These examples show the links between the teacher's mental models, represented by horizontal arrows in Figure 1.

The teacher's mental or espoused models of teaching and learning mathematics, subject to the constraints and contingencies of the school context, are transformed into classroom practices. These are the enacted (as opposed to espoused) model of teaching mathematics, the use of mathematics texts or materials, and the enacted (as opposed to espoused) model of learning mathematics. The espoused-enacted distinction is necessary because case studies have shown that there can be a great disparity between a teacher's espoused and enacted models of teaching and learning mathematics (for example, Cooney, 1985). Two key causes for the mismatch between beliefs and practices are as follows.

First of all, there is the powerful influence of the social context. This results from the expectations of others including students, parents, peers (fellow teachers) and superiors. It also results from the institutionalized curriculum: the adopted text or curricular scheme, the system of assessment and the overall national system of schooling. These sources lead the teacher to internalize a powerful set of constraints affecting the enactment of the models of teaching and learning mathematics. The socialization effect of the context is so powerful that despite having differing beliefs about mathematics and its teaching, teachers in the same school are often observed to adopt similar classroom practices.

Secondly, there is the teacher's level of consciousness of his or her own beliefs, and the extent to which the teacher reflects on his or her practice of teaching mathematics. Some of the key elements in the teacher's thinking — and its relationship to practice — are the following:

awareness of having adopted specific views and assumptions as to the nature of mathematics and its teaching and learning;
the ability to justify these views and assumptions;
awareness of the existence of viable alternatives;
context sensitivity in choosing and implementing situationally appropriate teaching and learning strategies in accordance with his or her own views and models;
reflexivity — being concerned to reconcile and integrate classroom practices with beliefs, and to reconcile conflicting beliefs themselves.

These elements of teachers' thinking are likely to be associated with some of the beliefs outlined above, at least in part. For example, the adoption of the role of facilitator in a problem-solving classroom requires reflection on the roles of the teacher and learner, on the context suitability of the model, and probably also on the match between beliefs and practices. The instrumental view and the associated models of teaching and learning, on the other hand, require little self-consciousness and reflection, or awareness of the existence of viable alternatives.

Mathematics teachers' beliefs have a powerful impact on the practice of teaching. During their transformation into practice two factors affect these beliefs: the constraints and opportunities of the social context of teaching, and the level of the teacher's thought. Higher level thought enables the teacher to reflect on the gap between beliefs and practice, and to narrow it. The autonomy of the mathematics teacher depends on all three factors: beliefs, social context, and level of thought. Beliefs can determine, for example, whether a mathematics text is used uncritically or not, one of the key indicators of autonomy. The social context clearly constrains the teacher's freedom of choice and action, restricting the ambit of the teacher's autonomy. Higher level thought, such as self-evaluation with regard to putting beliefs into practice, is a key element of autonomy in teaching. Only by considering all three factors can we begin to do justice to the complex notion of the autonomous mathematics teacher.

References

COCKCROFT, W.H. (1982) *Mathematics Counts*, London, HMSO.

COONEY, T.J. (1985) *Journal for Research in Mathematics Education*, 16, 5, pp. 324–36.

NATIONAL COUNCIL OF TEACHERS OF MATHEMATICS (1980) *Agenda for Action*, Reston, Va., NCTM.

THOM, R. (1973) in HOWSON, A.G. (Ed.), *Developments in Mathematical Education*, Cambridge, Cambridge University Press, pp. 194–209.

THOMPSON, A.G. (1984) *Educational Studies in Mathematics*, 15, pp. 105–27.

24 *Becoming a Mathematics Teacher: Grounds for Confidence?*

John Hayter

In a note in Marion Bird's book, *Generating Mathematical Activity in the Classroom*', Afzal Ahmed suggests that the Cockcroft Report was about confidence: 'Children's confidence, teachers' confidence and public confidence in mathematics teaching in schools' (Bird, 1983). To what extent are we entitled to be confident about the state of the mathematics teaching profession?

At a time of low teacher morale, recently renovated initial teacher education courses and continuing problems with recruitment to mathematics teaching, the theme of confidence seemed an appropriate one to use for considering entry, training and professional future for new mathematics teachers. Such a review is not being conducted without an awareness of my personal decision twenty-five years ago to enter teaching or of my experiences in teacher education over about half that time. Since the latter period has been almost exclusively concerned with postgraduate students, this may determine or at least colour some of the views expressed.

Assuming that many readers are mathematics educators at some point within the system, it is worth asking the question,'Would you, given your career choice again in 1989, choose to enter mathematics teaching?' Would our answers be affected chiefly by the experience of being a teacher, by the perceived nature and state of mathematics teaching today or by the current wider range of opportunities which are available to mathematics graduates? In practice it is probably impossible to give any kind of reliable answer, but the evidence of a falling proportion of mathematics graduates entering teaching suggests that of four of us deciding to teach in the early 1960s, only one would do so now. Does it matter? I would suggest that it is not irrelevant if one accepts that the nature of new entrants must be one determinant of the teaching profession. The training they are given and the receiving state of the profession must be the other major factors.

One further piece of self-examination is necessary before considering some aspects of confidence: as mathematics educators in initial training are we essentially recruiting officers for the system — a system with a high success rate (as judged by the proportion of students on courses who obtain a professional qualification) and a high placement rate (proportion of certificated persons who obtain a teaching post)? Or are we honest brokers playing a significant part between an education system needing teachers and young people making a career choice? Or more distantly perhaps, are we through AUMET, NATFHE and SCAMES,[1] members of pressure groups for raising levels of awareness and provoking action on quality and quantity? Or are we in practice operating at the baser level of perpetuating our own kind or even protecting our own future employment by ensuring that there are sufficient customers? Such issues seem worth consideration at a time when there is encouragement of various kinds to increase the flow of new teachers into mathematics teaching, and when teacher educators are significant in designing the schemes, attracting the customers and carrying out the training.

Confidence in the Recruits to Courses

Meeting aspirants to teaching at interview is a time-consuming but not uninteresting part of my work — particularly if for a significant proportion of the interviewees the meeting is a first tutorial. Interviewing with an experienced teacher is a luxury which I have welcomed in these days of cutbacks. The reaction of many such teachers is one of surprise at most applicants' lack of awareness of educational issues and a disappointment that so few are able to enthuse about their mathematical studies at university. Their application forms express enthusiasm and interest in mathematics but discussion usually reveals this to be linked to their success and their experience of mathematics at *school* level.

For 'mature' applicants the present aliveness of their mathematics can be a greater problem. A handful of Open University graduates have been notable exceptions. However, for many their statement that mathematics was their favourite subject at school twenty years ago projects an image of mathematics as a repository of knowledge rather than a current interest. Tentative inquiries about recent involvement with mathematics usually lead to variously dressed apologies for not doing any further (meaning even higher) mathematics. Yet I cannot believe that interviews relating to an aspiring English teacher's current interest/activities in their field would produce such an arid outcome. Interest in literature, the theatre, writing, could be expected to be current rather than just successfully encountered while at school; further one might expect current interests to be readily communicable.

In recent years it has been illuminating to require the mathematics students in my PGCE group to teach us something unrelated to mathematics. The ease of communication and particularly their enthusiasm contrast with much that is to be seen in later months in maths classrooms. Sharing knowledge and pleasure at making pastry, playing an instrument, climbing a rock face or communicating in a foreign language reveal characteristics which the teaching of mathematics all too easily seems to smother.

But why do people choose mathematics teaching as a career? Or perhaps more appropriately, who do people choose to enter a course in training to become a maths teacher? Straker (1984) inquired into the attitudes of final year mathematics undergraduates in six universities towards mathematics teaching. Of the 17 per cent of respondents who put teaching as first choice (likely to be an overestimate in relation to the whole graduate output from these universities) there were noticeable differences between women and men. The views of the female students suggested a higher level of commitment and caring than their male counterparts, whose responses reflected a greater concern about material issues.

In my experience it is often difficult to disentangle students' motivations. Some students who describe a long-term commitment to becoming a teacher can face as many surprises/crises as those who have decided negatively to go into teaching, *against* going into 'computing, accountancy and other "office" jobs'. A wish to do something useful in society (often seen as an application of their Christian faith) features prominently, together with a wish to recapture an enjoyed experience while a school pupil and not infrequently to imitate a respected teacher. Few appear to believe that educationalists can change the world, and regrettably many have little experience in working with children. Individual tuition of a brother or a friend may be their only experience of teaching but it has had a positive influence. Without Sunday school and church youth group experience the pool of experience of working with young people would be drastically reduced.

What seems likely is that in deciding to train as a secondary mathematics teacher a graduate may experience the response from press and general public that might be expected by someone joining a minority religious sect: surprise, lack of comprehension and a concern for future well-being.

Concerns about the shortage of mathematics teachers have reached new heights in the late 1980s. Special funding for innovative courses and bursaries for those undertaking PGCE courses reflect a growing awareness of a chronic problem in mathematics and several other subjects.

What Kind of Shortage?

In 1963 a subcommittee of The Mathematical Association reported on the supply and training of mathematics teachers. The introduction (Mathematical Association, 1963) states:

> The shortage of mathematics teachers is now generally recognized: the danger is that it may be accepted. Available statistics, however incomplete and however cosily interpreted, reveal serious shortages, some now chronic, all now seriously impairing quality. (p. 1)

Although over 40 per cent of honours graduates in 'mathematics' were entering teaching at that time (300 out of some 700 in 1961), half of these were entering without a professional training. In grammar schools alone it was estimated that the deficiency in honours graduates could absorb two years' total output of mathematics specialists. But at the same time expanding training colleges were said to be in need of about 250 mathematics graduate lectures in the following few years, and the supply of potential university teachers was said to be only half that required to maintain university staffs. Even for the relatively small number of mathematics posts in university Departments of Education there was a shortage of adequately qualified applicants.

To read this catalogue of shortages is to realize that shortage situations do change! During the 1980s unusually able young mathematics graduates have had very little opportunity to gain lecturing posts in university mathematics departments — a state of affairs which led Jones (1981) to suggest compulsory retirement of the members of the 35–45 'bulge' group to be found in university departments as the most effective solution.

The same paper reports an output of some 3000 mathematics graduates per annum from universities in the late 1970s. In addition, there were several hundred from polytechnics. At that time about 12 per cent of mathematics graduates were going into teaching. The prospect of declining school populations offered hope for a significant improvement in the supply of mathematics teachers in the mid to late 1980s, before the expected decline in the number of entrants to universities in the early 1990s has its effect on output just at the time of an increasing secondary population. This future, as opposed to present, problem is illustrated in the DES's consultative document, *Action on Teacher Supply in Mathematics, Physics and Technology* (DES, 1986; see Table 1).

As the consultative document points out, it is useful in discussing the shortage of mathematics teachers to distinguish between overt shortages (unfilled vacancies), hidden shortages (tuition given by inadequately qualified teachers), and suppressed shortages (subject under-represented on timetable because of a lack of suitable teachers). The first is easiest to identify while the last is perhaps of less significance in mathematics than it is in physics and CDT. However, the teaching of mathematics by those with minimal qualification is a particular problem in a subject which all teachers have studied in secondary school and many have carried as a potentially useful subsidiary subject.

Table 1.

	Secondary school population (thousands)	*Population aged 20–24 (thousands)*
1980	4110	3630
1985	3750	4140
1990 (Projected)	3020	3950
1995 (Projected)	3130	3370

Source: DES, *Action on Teacher Supply in Mathematics Physics and Technology*, London, DES, 1986.

The statistics for shortage are notoriously difficult to collect, but the figures given in DES (1986) are likely to be as reliable and up-to-date a guide as any we could expect. What is not easily apparent is the progress of particular cohorts of applicants from the number applying to the number becoming established members of the profession. The use of statistics from DES (1986), GTTR and UCET[2] enables some picture of the process to emerge, though at times the degree of correspondence between data from these sources is difficult to establish.

From Applicant to Qualified Teacher Status

Various routes towards becoming a qualified teacher in secondary mathematics are possible. The routes and size of enrolment for courses starting in 1985 are:

Postgraduate Certificate	through GTTR	666	(universities 529 polytechnics and colleges of higher education 137)
	non GTTR	30	
Concurrent BSc or Honours BEd		160	

Statistics prepared by the GTTR make it relatively easy to monitor the entry to postgraduate courses, while this source and data collected by UCET make it possible to follow the intake to university departments through to their eventual post-course employment.
Using the 1985 entry data again we have:

Total number of applications for PGCE mathematics places	1182
Number of these who enrolled for the course in October 1985	666
Number withdrawing from the GTTR system at some stage	458
Number who remain in the GTTR system but were not placed	58

Those withdrawing fall into three categories

- those who had received an offer of a place (estimated to be about 30 per cent of withdrawals);
- those who withdrew before receiving an offer;
- those applicants assumed to have withdrawn after failing to respond to GTTR offers to try further institutions.

Of the 529 on university courses 477 were successful in obtaining a PGCE qualification; 390 obtained teaching posts in the UK; fifteen were still seeking teaching jobs in October 1986.

These figures are somewhat smaller than those for the previous year, but proportionally similar. They show approximately 10 per cent of entrants not successful completing the course, rather less than 20 per cent not entering teaching on successful completion and the remaining 70 per cent of the entry obtaining a teaching post.

To observe current patterns is not to give them the status of having any particular long-term significance. It is possible that we might accept or reject a higher proportion of applicants or fail a higher proportion of students following a course. Nevertheless, when numbers of applicants and numbers of entrants are often quoted independently, it is worthy of note that the total applications for PGCE places of about 1200 became an entry of about 500 teachers (1985–86 figures for all PGCE courses) in a system with a target of rather more than 900 training places.

A Local Problem?

At the Fifth International Congress on Mathematical Education held in Adelaide in 1984 one working group considered the problem of shortage of mathematics teachers. Australia, America and France were amongst those reporting shortages. Many of the reasons were familiar: poor status and unattractive conditions of teaching, more lucrative posts in commerce and industry particularly in relation to computers. In Australia postings to bush schools in the early years of teaching were unattractive. In France, where the problem is more recent and regional, it was suggested that the dogmatic teaching style which has been prevelant in French schools tended to turn students against mathematics (Carss, 1986, p. 93).

Yet the problem is not universal. In Malaysia excess mathematics teachers are currently being retrained for teaching English as a second language. Closer to home it was reported in late 1986 that the Secretary of State for Education was considering the possibility of importing some of the surplus German mathematics teachers to ease our problems. The roots of such a surplus are presumably embedded in a whole social structure of attitudes, expectations, opportunities and rewards in the worlds of commerce and of education. No simple technique seems likely to produce a radical change in the situation.

Confidence in Quality of Entry

People outside the teacher education field sometimes ask me about the quality of entry in these times of shortage. At meetings of teacher educators I sometimes hear claims that an institution is maintaining its standards of entry in spite of a dearth of applicants. 'Am I?', I wonder. Judgments about teaching potential on the basis of an application form, references and an interview always seem unreliable, even if a practising teacher is involved. One is always taking risks in selection, and at times of shortage more risks are likely to be taken than in times of plenty. Mathematics students do fail PGCE courses — perhaps more than do students in other subjects. Bishop and Nickson (1983, p. 41) suggest that there is evidence that more mathematics teachers than teachers of other subjects have their probationary year extended.

But absolute quality is difficult to assess. The list of desirable qualities for a teacher suggested in various DES and HMI publications often seems somewhat unrealistic in the light of the total population of applicants being considered. Energy, resilience and marked communication skills, while undoubtedly welcome, do not always appear in abundance at interview, yet some entrants certainly demonstrate this range of qualities as the year progresses. One measure of academic quality which is available is that of class of degree. While it is probable that the correlation between class of degree and teaching quality is modest, it is noticeable that in university Departments of Education the quality of entry by degree class is markedly different for the mathematics entrants than for the entry population as a whole. It is even more marked if we compare mathematicians and historians. Data for the 1986 entry to university Departments of Education, collated by UCET, for the percentages of group intake under degree class are shown in Table 2.

So while less than 2 per cent of history entrants have lower than a second class degree, about 40 per cent of mathematics students have this background of achievement in their first degree. We might find ourselves agreeing with the teacher who commented that if there are a limited number of mathematics graduates around, she would rather that it was the able ones designing the planes and guided missiles. My concerns are two-fold. First, what sort of an educational/mathematical experience was it in getting a third class or pass degree? By definition, it seems to suggest very modest success when it is borne in mind that very few students fail their degree altogether. Second, what are the longer-term implications for the profession in decision-making about the curriculum, in A-level teaching, etc.? Will it be possible in any sense to expect a different quality

Table 2.

	Higher degree	*1st*	*2i*	*2ii*	*2undivided*	*3*	*Pass/ general*	*Other*
Overall	3.0	2.9	33.2	39.2	2.5	9.4	7.3	2.5
History	3.3	3.3	47.2	40.7	3.9	0.7	1.0	—
Maths	2.8	4.3	19.3	27.9	5.2	19.7	17.8	3.0

Source: University Council for the Education of Teachers, UCET.

in history and mathematics education in schools because of this difference in academic quality of entry? Do we consciously or unconsciously modify what we do as teacher trainers in the light of this difference? *If* it is irrelevant, should we be campaigning amongst our colleagues for historians with third class and pass degrees to be given a chance as teachers?

Dore's book, *The Diploma Disease* (Dore, 1976), made evident on a world scale the nonsense and the dangers of chasing inappropriate qualifications. A PhD is not necessary to drive a taxi, but it might help in some countries if it puts you ahead of the rest of the field in formal qualifications. As a disappointed applicant for a taxi driving post, one might have to revise one's academic aspirations if a PhD becomes necessary in the scramble for jobs. This educational inflation is usually evident where jobs are in short supply, but there are many jobs in which people feel overtrained for what the job actually requires them to do. Some have even suggested this about the relatively prestigious job of general practitioner, for which the training is some eight years post-A-level. Is it also true for the mathematics teacher?

The most recent messages from HMIs and the DES are unclear. On the one hand, training institutions are being encouraged to accept only those with a strong academic background in mathematics, while on the other, innovative methods are being sought to bring in, through retraining or longer PGCE courses, students whose background in mathematics is considerably weaker. While many of us might argue that a mathematics course focused on the needs of the intending teacher may be more valuable than much of what is experienced and 'failed' by many a mathematics graduate who enters a PGCE course, it does raise questions about the knowledge needed to be a mathematics teacher and the relative importance of other qualities which are not subject specific.

Beyond Initial Training

Longitudinal studies of teachers of mathematics seem to be in short supply. Yet concerns over input and output are short-sighted unless viewed against longer-term contributions to teaching. Professional life expectancy is of significance both to course designers and tutors and to those financing such training programmes. A one-year programme, even of thirty-six weeks, may be all too short for preparing someone for the demands of full-time teaching, but it is an expensive investment if the teacher only contributes for two or three years in the classroom.

A pilot study of teacher movement carried out by Cornelius (1981) in sixty-nine secondary schools in the north-east of England recorded 127 changes of mathematics staff in a two-year period (1978–80). While forty-seven involved a change of school and twelve were through retirement or death, some thirty-four represented true wastage by virtue of a move into another job, failure or illness/personal reasons. The future of the remaining thirty-four who moved for reasons of pregnancy, family, movement of spouse was uncertain with respect to possible re-entry to mathematics teaching.

More recent data in DES (1986) suggest wastage figures of 8–10 per cent per annum. It seems likely (and Cornelius' pilot study figures given support) that wastage rates are likely to be higher in the early years of teaching as people decide whether teaching is an appropriate career for them. Adding to this the other hazards (pregnancy, spouse movement without necessarily another teaching job being found), the loss in the early years could well be 15 per cent per annum. If we use this model, the number of our original 500 entrants surviving to years 2–5 would be 425, 360, 305, 260. Such statistics, even if only crude projections, raise major questions about counteracting wastage, the timing of training investment and the ability, achievement and characteristics profile of those who survive.

Data and analysis of inadequacies can present a rather depressing view of what is, as opposed to what might have been. Yet the experience of working with those in training is probably as rewarding as it ever was. The talents, skills and initiative of some are a frequent palliative to other indicators of quality. Development in PGCE courses give cause for optimism about the linking of theory and practice and the relationship between teachers and teacher educators. Only the complexity of the teacher's work in the late 1980s makes one query whether we are moving fast enough to make the preparation adequate.

Finally, what confidence is there in the receiving profession as new entrants join? There can have been few more difficult times to join the teaching profession than in the last year or two, with the low state of morale. Yet the support systems for new teachers are probably better than ever before: schools with only limited numbers of probationers and consequently more scope to look after them; heads of department with clearer ideas on their role in looking after a new probationer; advisory teachers largely freed from administrative chores to be able to work alongside new teachers in schools; two professional associations more active than ever in providing materials and in organizing meetings and conferences. Amid the prevalent gloom there are signs to encourage.

Notes

1 AUMET — Association of University Mathematics Education Teachers; NATFHE — National Association of Teachers in Further and Higher Education; SCAMES — Standing Committee of Associations concerned with Mathematics Education in Schools.
2 GTTR — Graduate Teacher Training Registry; UCET — University Council for the Education of Teachers.

References

BIRD, M. (1983) *Generating Mathematical Activity in the Classroom*, Bognor Regis, West Sussex Institute of Higher Education.

BISHOP, A. and NICKSON, M. (1983) *Research on the Social Context of Mathematics Education*, Windsor, NFER-Nelson.

CARSS, M. (Ed.) (1986) *Proceedings of the Fifth International Congress on Mathematical Education*, Boston, Mass. Birkhauser.

CORNELIUS, M. (1981) 'The Wastage and Movement of Secondary Mathematics Teachers', *The Bulletin of the Institute of Mathematics and Its Applications*, 17, July.

DEPARTMENT OF EDUCATION AND SCIENCE (1986) *Action on Teacher Supply in Mathematics, Physics and Technology*, London, DES.

DORE, R. (1976) *The Diploma Disease*, London, Unwin.

JONES, D.S. (1981) *Whither Mathematics*, Paper prepared for and made available by University Grants Committee.

MATHEMATICAL ASSOCIATION (1963) *The Supply and Training of Teachers of Mathematics*, Leicester, The Mathematical Association.

STRAKER, N. (1984) 'An Unattractive Option?', *The Times Educational Supplement*, 11 May.

Notes on Contributors

Neville Bennett is professor of primary education at the University of Exeter

David Burghes is professor of education at the University of Exeter, where he directs the Centre for Innovation in Mathematics Teaching

Leone Burton is professor of mathematics education at the Thames Polytechnic, London

Charles Desforges is professor of primary education at the University of Exeter

Paul Ernest is a lecturer in education at the University of Exeter

Jeff Evans is a senior lecturer in statistics at the Middlesex Polytechnic, London

Kath Hart is professor of mathematics education at King's College, London, where she directs Nuffield Secondary Mathematics

John Hayter is a senior lecturer in education at the University of Bristol

David Hobbs is a senior lecturer in education at the University of Exeter

Zelda Isaacson is a senior lecturer in mathematics education at the Polytechnic of North London

Barbara Jaworski is a lecturer in mathematics education at the Open University

Dudley Kennett is a lecturer in education at the University of Exeter

Dietmar Küchemann is a lecturer in education at the University of London Institute of Education

Stephen Lerman is a senior lecturer in mathematics education at the Polytechnic of the South Bank, London

John Mason is a senior lecturer in mathematics education at the Open University

Jenny Maxwell is a former numeracy tutor at the WELD Community Education and Arts Project, Handsworth, Birmingham

Marilyn Nickson is a principal lecturer in mathematics education at the Essex Institute of Higher Education

Susan Pirie is a lecturer in mathematics education at the University of Warwick

Derek Stander is a former teacher of mathematics at Burleigh County Secondary School, Plymouth

Derek Woodrow is a principal lecturer in mathematics education at the Manchester Polytechnic

Index